THE COMPLETE BOOK OF
PLANT PROPAGATION

THE COMPLETE BOOK OF
PLANT PROPAGATION

Robert C.M.Wright and Alan Titchmarsh

A practical guide to the various methods of propagating trees,
shrubs, herbaceous plants, fruits and vegetables

Ward Lock Limited·London

© Ward Lock Limited 1981

This edition first published in Great Britain in 1987
by Ward Lock Limited, 8 Clifford Street,
London W1X 1RB, an Egmont Company.

Reprinted 1988

House editors Denis Ingram and Deborah Maby
Designed by Heather Sherratt

Cover photographs
Front: Alan Spain
Back: Bob Challinor

Text filmset in 10 on 11 point VIP Plantin
by Facet Filmsetting Limited, Southend-on-Sea.

Printed and bound in Great Britain by
Hollen Street Press Ltd, Slough, Berkshire

British Library Cataloguing in Publication Data

Wright, Robert Cecil Murray
    A handful of plant propagation.–
    Revised ed.
    1. Plant propagation
    I Title
    635'.04 3   SB119

ISBN 0-7063-6412-0

Contents

Preface 7

Acknowledgements 9

1   The Principles of Propagation 11

2   Facilities and Equipment for Propagation 19

3   Hybridization and Plant Inheritance 27

4   Harvesting and Storage of Seed 32

5   The Process of Germination 37

6   Seed Sowing in the Open 42

7   Seed Sowing Under Glass 51

8   Natural Vegetative Increase 60

9   Artificial Methods of Vegetative Increase 66

10   Mist Propagation 69

11   New Plants from Cuttings 73

12   The Techniques of Grafting and Budding 85

13   Layering Plants 99

14   The Propagation of Trees and Shrubs 105

15   How Roses are Increased 119

16   Propagation of Herbaceous Perennials and Water Plants 122

17   Multiplication of Alpines 129

18   Increasing Bulbous Plants 136

19   Raising Annuals and Biennials 141

20   The Propagation of Hardy Fruit 146

21   Raising Vegetable Crops and Culinary Herbs 154

22   How Decorative and House Plants are Increased 159

23   Plant Pests and Diseases in Relation to Propagation 167

Index 171

Preface

Growing plants is a satisfying craft and one which, for all its pitfalls, becomes increasingly absorbing. Sooner or later any gardener may be forced to produce his own plants, for economic reasons if nothing else, and it is then that the thrill of propagation will become apparent.

No complicated equipment is necessary to produce a dozen new shrubs from an existing one, or a few trays of bedding plants from a packet of seeds. On a more ambitious scale, if the enthusiast wants to turn part of a greenhouse into a plant factory then apparatus is available to make life easier and propagation even more productive.

This book sets out to show the techniques of propagation that can be used by anyone with a modicum of interest and an ounce or two of determination. The raising of plants from seeds and cuttings, by layering, budding and grafting are all described in detail. Line drawings are used freely to make the operations clear to the novice. There is advice, too, on installing propagating frames and other equipment, including a section on mist propagation revised by one of the country's leading pioneers, Mr. Douglas Lowndes, in this time-saving method of plant production.

Originally published in 1955 under the title *Plant Propagation*, this book was highly commended by the gardening press, and important advances in propagation techniques resulted in its being revised in 1973. This latest edition has once again been completely revised to include much more information on up-to-date techniques and modern equipment. Added to this, over a hundred new plant entries have been incorporated to increase the book's value to the gardener anxious to find a reference work that will tell him exactly how to propagate a given plant.

This new edition includes information on the pregermination and fluid sowing of seeds, up-to-date methods of weed control between young plants, and many modern aids that cut down the risks involved in plant propagation.

Whether it is your intention to grow vegetables and flowers from seeds, to propagate your own roses and fruit trees by budding and grafting, or to multiply your plants without the aid of a greenhouse, you will find all the information you need between the covers of this book.

You may start to propagate plants as a means of saving money, but it will not be long before you discover that propagation is one of the gardener's richest pleasures.

R.C.M.W.
A.T.

Acknowledgements

The publishers gratefully acknowledge Fisons Ltd for granting permission to reproduce the colour photograph (Plate 1a) on p. 97, and Halls Homes and Gardens Ltd in providing the location for the equipment shown in Plates 1b and 1c.

The following line drawings are by Nils Solberg: Figs. 4–8, 10–14, 16–26, 28, 33–34, 36, 39–42, 45–47, 49–61 and 63–68.

Fig. 4a and b are after Figs. 41 and 39 respectively in *Textbook of Botany*, Lowson, published by University Tutorial Press, 1945; Figs. 5, 18, 23, 24, 25 and 34 are after the illustrations on pp. 6, 21, 20, 30, 25, and 10 respectively in *The Humex Book of Propagation*, J. Harris, published by Macdonald, 1980; Figs. 6a and 21 are after the illustrations on pp. 11 and 23 in *Electricity in your Garden*, published by The Electricity Council, 1976; Fig. 6c is after an illustration kindly supplied by Eltex Ltd; Fig. 8a is after illustration kindly supplied by Jemp Engineering Ltd; Figs 8b and 26 are after the illustrations on pp. 23 and 39 respectively in *Gardening Under Cover*, A.

Titchmarsh, published by Hamlyn, 1979; Figs. 12, 13 and 56 are after illustrations on pp. 31, 32 and 87 respectively in *Plant Propagation*, P. McM. Browse, published by Mitchell Beazley, 1979; Fig. 20a and b are after illustrations on pp. 28 and 288 respectively in *New Illustrated Guide to Gardening*, published by Reader's Digest Association Ltd., 1979; Fig. 22 is after illustrations by Fisons Ltd; Figs. 46 and 65 are after illustrations on pp. 44 and 23 respectively in *Amateur Gardening Plant Propagation*, ed. A. G. L. Hellyer, published by W. H. Collingridge Ltd, 1955; Figs. 47, 49, 50, 51, 52, 53, 54a and b, 55a and b, 57, 59 and 60 are after Figs. 57, 107, 63, 66, 49, 52, 78, 80, 110, 113, 33, 42, and 46 respectively in *The Grafter's Handbook*, R. J. Garner, published by Faber and Faber Ltd, 1970; and Figs. 63 and 64 are after Figs. 2 and 1 respectively in *Fruit Tree Raising*, Bulletin 135, published by H.M.S.O., 1969. The publishers regret that, despite considerable effort, they have been unable to identify the copyright holder of an illustration, a modification of which is depicted in Fig. 41b, to whom acknowledgement is made.

1
The Principles of Propagation

Plant characteristics that affect propagation

Even the casual observer cannot fail to be impressed by the amazing diversity of plant life. Plants are found varying in size and in shape, in form and in habit, while modifications in structural details are endless. This variation has largely been brought about by the ceaseless struggle for existence that occurs everywhere in nature. Therefore, many of the modifications or adaptations have considerable bearing on a plant's perpetuation and increase. Familiar examples of this are the runners of the strawberry, the tubers of the potato and the rhizomes of the German iris, all of which are modified stems adapted to serve as a natural means of reproduction.

Adaptations in plants often relate to seed propagation which is a common natural method. The bright colours of many flowers and the fragrance of others are designed to attract insects, so that they may carry out the vital task of pollination. The nectar is the bees' further inducement to attempt this operation which, if accomplished successfully, usually results in fertilization and seed.

Nature also takes a hand at seed sowing, and many ingenious devices in plants are employed to facilitate the dispersal or scattering of seed by wind and other means (Fig. 1).

Plant cultivators throughout the ages must have learnt a great deal about propagation by observing nature's methods and by trying to imitate them. For instance, a general disadvantage of natural vegetative increase, as opposed to sexual reproduction, is that the new individuals are usually crowded together, resulting in severe

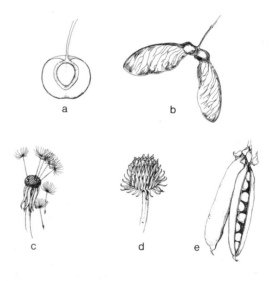

Fig. 1. Different methods of seed distribution. (*a*) Typical stone fruit: when this falls to the ground the fruit rots and provides nutrients for the germinating plant. (*b*) Sycamore seeds have wings which ensure that they will be distributed some distance away from the parent tree. (*c*) Dandelions have seeds on top of which are parachute type structures designed to enable the wind to disperse the seeds over great distances. (*d*) Burdock, is a typical burr fruit: it sticks to an animal's fur, is carried by the animal and then rubbed off so that it falls to the ground some distance from the parent plant. (*e*) Some leguminous plants distribute their seeds by means of an exploding pod.

competition for nutrients and light. But the gardener can lend a hand here by separating such natural creations as bulbs, corms and tubers, and by spacing them out in the soil.

Occasionally, division of plants occurs naturally. An instance is the rooted portion of a brookside plant torn from the main plant by the flow of the stream and carried to a new position where it may become established.

Several methods of plant increase are described as artificial, but it is claimed that with the exception of budding all of them occur in nature. The twigs of the crack willow *Salix fragilis* are snapped off easily by wind, or by birds alighting on them. Should these broken stems come to rest on soft ground they take root readily. Branches of certain cacti sometimes fall on the soil and have been observed then to produce roots. Similarly, the leaves of such a plant as the sedum often take root when accidentally knocked off the plant. Perhaps it was instances like these that first indicated how plants could be propagated from cuttings of stems and leaves.

Pieces of the roots of several plants such as the false acacia (*Robinia*) and the sumach (*Rhus*) are often used for propagation. This may have been discovered when suckers or shoots were noticed arising naturally from the roots of such plants. Perhaps the secret was revealed when a plant of this type was blown over by the wind. Nature uses 'layering' as a means of plant increase quite frequently. Thus, the branch of a tree sometimes breaks or bends down by its own weight or by pressure from wind so that it rests on the ground. If it remains attached to the tree it continues to grow and where it touches the soil rooting may occur. Another example of natural layering is the blackberry whose canes take root when their tips come in contact with the soil, resulting in the production of new plants. Two other forms of asexual reproduction are shown in Fig. 2.

Grafting occurs without man's intervention, but in such circumstances is purely accidental and plays no part in natural plant increase. A good example is the common ivy whose stems often unite with one another where they meet, forming natural grafts. Similarly, branches of trees often become joined together where two of them make accidental contact.

Knowledge of plants helps

These examples show that much can be learnt

Fig. 2. Vegetative or asexual reproduction. (*a*) In *Chlorophytum*; (*b*) Creeping buttercup increases asexually by runners which root at frequent intervals.

about propagation from nature, yet with many plants experience alone teaches us the most reliable method. The study of the plants' internal structure and how it works has thrown some light on propagation. In this way botanists have often been able to advise the practical propagator. Today, systematic investigation in this and other aspects of plant culture is undertaken by horticultural research stations in various parts of the world. New and improved methods are constantly being discovered.

Botanists have divided most flowering plants into two large groups, or classes, and from a practical point of view these are not difficult to distinguish. One class, the monocotyledons, comprises mainly grasslike plants and bulbous rooted species such as grasses, onions, daffodils, gladioli and montbretia. All other ordinary flowering plants including tree and shrubs (apart from conifers which form a special group of their own) are dicotyledons. The different ways by which monocotyledons and dicotyledons germinate are shown in Fig. 3.

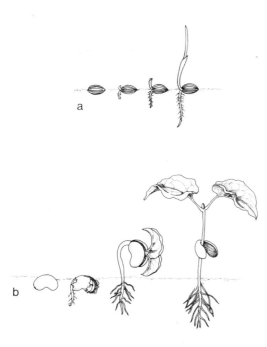

Fig. 3. Germination of (*a*) monocotyledonous and (*b*) dicotyledonous seedlings. The monocotyledons produce only a single cotyledon (seed leaf), whereas a dicotyledon usually produces two seed leaves.

There is one very significant distinction between monocotyledons and dicotyledons which has an important bearing on propagation. This is the presence of a tissue called cambium in the roots and stems of dicotyledons and of coniferous plants, which is not found in the monocotyledons. Cambium consists of a layer of actively dividing cells. It is the growing area of stems and roots, and causes their annual increase in girth.

Cambium plays a vital role in propagation. In cuttings of various kinds it gives rise to the roots they produce. It is the actively dividing cells of the cambium that enable a union to be formed between portions of different plants as occurs in grafting and budding. The cambium is not conspicuous, so a knowledge of its exact position in stems and roots is important to the propagator. It is found just beneath the bark or rind of woody plants.

The grass-like monocotyledons do not possess a cambium layer and cannot, therefore, be increased by cuttings, grafts or layers. An exception to this rule is a plant called dracaena which, although belonging to this class, has cambium tissue and may be increased by cuttings or layering.

In relation to their span of life plants are placed in three groups called annuals, biennials and perennials.

Annuals are plants which complete their life cycle within 12 months. They grow from seeds, flower, produce seeds and die within one growing season. Examples are peas and beans, and the flowering annuals such as larkspur and clarkia.

Biennials normally live for two seasons or parts of two. In the first season the plant grows and builds up a food reserve which is stored during the winter while the plant is dormant. The second season is devoted mainly to the production of flowers and seed from the materials stored. Many common vegetables such as turnips and parsnips are biennials, while Canterbury bells and honesty are examples of biennial ornamental plants.

Perennials live for an indefinite period and are capable of storing food year after year and using it normally for the annual production of seed. There are two types, herbaceous perennials and woody perennials. The former produce shoots above ground of a succulent nature which usually die down at the end of the growing season. The underground parts survive and growth is renewed each spring. Potatoes, delphiniums and many bulbous plants are examples.

Trees and shrubs are the woody perennials. In this climate many of them lose their leaves with the approach of winter and remain dormant during the cold period. Their permanent framework of stems extends in growth each season.

This classification applies in a general sense, but, from the gardener's point of view, is not by any means inflexible. Thus some perennials succeed best when treated as annuals or biennials, examples being wallflowers and sweet Williams. There is also experimental evidence that the life of certain annuals and biennials can be extended indefinitely by simply preventing them from flowering.

In ordinary garden practice the increase of annuals and biennials by any portion of their vegetative parts is rarely advantageous. Seed is the only practical means of increase. Perennials are multiplied both by seed and from portions of their stems, leaves and roots. The most convenient and suitable method depends upon the particular plant.

Some complexities

Stem cuttings provide a ready means of increasing a wide range of plants, but there is a marked difference in the ease with which various kinds may be rooted. As a general rule cuttings from plants with a lot of pith in their stems do not strike readily and, on the other hand, trees and shrubs with extremely hard wood, like holly, are also difficult or impossible.

Usually, however, one can only say from experience whether a given plant will root easily or not. Blackcurrants root readily in ordinary soil in the open, but under similar conditions cuttings from apples and pears will rarely strike, Certain rhododendrons can be increased from stem cuttings, but with a number of other species and varieties of this genus cuttings may fail to root. Mist propagation, however, has greatly extended the range of rhododendrons and other plants which can be increased from cuttings.

Similarly with grafting; although the majority of woody plants may be increased in this way a few such as walnut and hickories are very difficult to graft successfully. Other complications relating to propagation by cuttings or grafts occasionally arise. For instance, if the lateral shoots of certain conifers such as species of abies, picea and araucaria are rooted or grafted, they grow readily but fail to develop into normal plants. Instead they remain permanently as flat-shaped typical laterals with dwarf straggly growth. However, if the main stem is used as a cutting, a normal cylindrical-shaped tree is produced.

When certain conifers such as *Juniperus virginiana* are raised from seed their foliage, habit of branching and type of growth are characteristic for several years. This is called the juvenile stage. Later there is a pronounced change in the characters mentioned and the mature plant assumes an entirely different appearance. If cuttings or grafts are taken at the juvenile phase of growth the resulting plants are of similar type. Although these may change to the mature form when the original seedlings do so, sometimes they remain permanently as typical 'juveniles'. Actually plants raised in this way have been distinguished as new horticultural varieties, well-known examples being *Chamaecyparis obtusa* and *C. pisifera*, sometimes known as retinosporas.

The degree of maturity of the parent plant is also known to affect the development of cuttings from certain other genera. For example, when cuttings are taken from mature specimens of begonias and gardenias the plants produced flower earlier than those propagated from immature individuals. In the case of carnations, cuttings secured from the terminal portions of vigorous young shoots with short internodes give the best results. Cuttings from older shoots have developed long internodes and usually give rise to weak straggly plants.

Hardiness affects propagation

Garden plants have been derived from almost every imaginable kind of climate. Some have come to us from the tropics and the deserts, or far up on mountain slopes in various parts of the world. Others have had their origin in these islands or have been secured from lands whose climate is generally warmer than that of Britain. It is not surprising, therefore, to find considerable variation in the climatic requirements of different plants, particularly with regard to temperature.

Plants which succeed in the open for the whole of their lives are termed 'hardy'. Another group are referred to as 'half-hardy' because they can only be grown unprotected in the open during the warmest months of the year. Examples are vegetable marrows and tomatoes. A third group called 'tender' plants must have some form of protection, such as a greenhouse, artificially heated in winter, for the whole of their lives. Cucumbers and gloxinias are examples.

Hardiness in plants affects propagation. Thus, many hardy plants may be raised in the open but, for reasons that will be discussed later, it is often an advantage to give some protection in the early stages of growth. Invariably half-hardy and tender plants are started under glass whether from seed or cuttings.

Half-hardy perennials are usually brought under cover in the autumn and protected from frost during the winter. In the spring these plants are, (a) planted directly outside, e.g. potatoes and gladioli, (b) started in heat and transferred outside later, e.g. tuberous-rooted begonias, or (c) allowed a temperature sufficiently high to induce the production of young shoots from the root stock to provide cuttings for propagation, e.g. dahlias and chrysanthemums.

Certain half-hardy perennials grow easily from seed and quickly develop into mature plants. It is not, therefore, considered worthwhile retaining them over the winter. Instead, they are treated as annuals and are raised every spring from seed. Examples of such plants are tomatoes, runner beans, antirrhinums and petunias.

Biennials which are required for seed production must be retained over the winter following their first season of growth. Those which are half-hardy are usually lifted in the autumn, are protected during the winter and are replanted outside in the spring.

When plants for cultivation in the open are raised in an artificially heated greenhouse or frame, either from seeds or cuttings, they must become gradually accustomed to open-air conditions before being moved outside. This process is called hardening-off. It generally consists of steadily lowering the temperature on successive days and gradually providing freer ventilation for a few weeks prior to transferring them outside.

Comparing seed and vegetative propagation

There are two distinct methods of plant increase:

(a) Seed (sexual)

(b) Vegetative (asexual)

Each of these methods has its advantages and limitations, but the one is complementary to the other. In some cases seed is preferable or is necessary, in other circumstances vegetative increase is advantageous or essential.

Seed

A seed may be defined as an embryo plant in a dormant condition and only requiring the right conditions to stimulate it into growth and development. The chief characteristics of a monocotyledon and dicotyledon seed are shown in Fig. 4. The origin of seed is due to a sexual function in plants which is involved in the process of pollination and fertilization. A seed may originate from one plant, but frequently there are two parents.

Owing to their mixed ancestry, seeds often give rise to plants that are neither exactly similar to their parents nor to one another. These differences may apply to the habit and vigour of the plant, the colour of the flowers, quality of the fruit and other important characteristics. This degree of uncertainty varies according to circumstances. For instance, the innumerable garden varieties of perennials such as 'Victoria' plums, the apple 'Ellison's Orange', 'Lloyd George' raspberry and varieties of lupins, delphiniums,

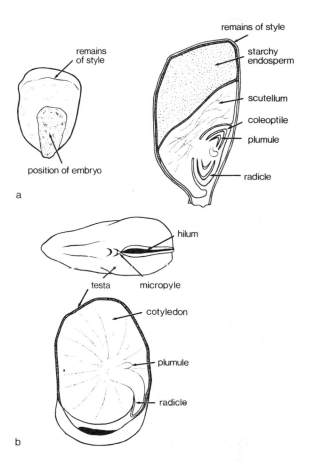

Fig. 4. Structure of seeds. (a) Longitudinal section of maize (monocotyledon) 'seed', which is really the fruit of *Zea mays*. The embryo, comprising plumule, radicle and scutellum, lies to one side of the mass of starchy endosperm. (b) External view and longitudinal section of broad bean seed (dicotyledon). The hilum is the point where the seed has broken away from its stalk within the pod. The embryo comprises a radicle, a plumule and two cotyledons.

chrysanthemums and roses, are all 'unfixed' hybrids. Seed from any of these will almost certainly result in a miscellaneous collection of plants, all of them probably different from and unfortunately usually inferior to their parents.

However, in the case of plants normally propagated from seed, such as most of our vegetables and annual and biennial flowering plants, the character of each variety has been 'fixed' by many years of selection, so that a good degree of uniformity and conformity to type has been acquired. Even with these there is some variation, and if the seed parents are selected indiscriminately, then deterioration of the stock is likely.

As distinct from hybrids there are many plants found in our gardens which are true species. These plants may be reproduced from seed with scarcely any noticeable difference or variation in type.

While raising plants from seed may result in variation and inferior offspring, conversely improved types and new varieties usually originate as seedlings. The process, known as 'selection', consists of saving the seed of any plant that shows an improvement, often quite slight, over the normal strain and continuing to select, often over many years, specimens that continue to show the improvement trend. More often new varieties are produced by artificial crossing or hybridizing, as it is usually called. This involves selecting two plants as parents and transferring pollen from a flower of one to a flower of the other under controlled conditions. The object is to combine in a new individual the best qualities of both the selected specimens.

In nature, seed is used extensively to increase plants, and often ensures the continuity of a species through difficult periods, like the winters of cool temperate climates. Thus most seeds are very resistant to severe cold and some, such as sunflower, pea and marrow seeds, have been known to survive several hundred degrees of frost under experimental conditions.

Seed is also an excellent natural means of promoting the wide dispersal of a species, thereby reducing competition between plants of the same kind. Man, too, has found seed very convenient for distributing plants, not merely in one district or country, but all over the world. A seed has been appropriately described as 'a plant packed and ready for transport'. Moreover, seed is easy to store, takes up very little room and can be kept for quite long periods, especially at low temperatures. Explorers who specialize in collecting plants from remote countries naturally prefer seeds as a means of sending the specimens they secure to other countries. Some of our best introductions have arrived in this form.

Seeds often afford a cheap and convenient method of raising large numbers of plants. Such plants are usually healthy, for there is less risk of transmitting diseases from parents to offspring than when vegetative methods are employed.

As always there are one or two exceptions to this rule. Seeds may transmit a number of serious diseases such as tomato mosaic, lettuce mosaic, celery leaf spot and halo blight of French and runner beans.

Another advantage of seed is that seedlings are often more robust than their parents. This applies particularly to hybrid seedlings which result after crossing two selected parents. This new race is known as the F_1 generation. With trees it is claimed that seedlings produce good deep anchoring roots superior to those of vegetatively increased plants.

A definite disadvantage of seed propagation for a number of species is that their seeds are difficult to germinate and may remain dormant for months or even years. Some kinds may not appear for at least a year and there are records of four years having elapsed before germination has occurred. During the waiting period such seeds must be allowed space and there is a risk of loss due to pests like mice and rats. There are various methods of overcoming or reducing the period of dormancy in seeds which will be discussed later.

Another fault of seed-raised plants is slowness to reach maturity. Seedlings of this type may require room in the garden and nursing for several years before they start to pay their keep. This applies to the majority of bulbous plants, while many of our fruit trees, if raised from seed, may take up to twenty years before producing flowers and fruit.

Slowness to reach maturity may affect propagation in another sense. Seedling trees such as beech, oak and chestnut may not flower until they reach the age of forty, while fir may take seventy years before producing seed. This means that should a new species of this type be introduced from abroad as a seed or seedling, no seed for further increase would be available for a generation or so. Obviously, in such circumstances vegetative methods of increase should be used if possible.

In some cases, seed offers the only means of increase, as with all true annuals and biennials. Conversely, certain plants fail to produce seed at all and must be multiplied vegetatively; horseradish and Jerusalem artichokes are examples. Certain species, too, produce flowers having male organs only on certain plants, while the female flowers are borne on separate individuals. This means that either the two distinct types of plants must be adjacent or a composite 'male and female' plant be secured by grafting or budding, before seed can be produced. Species which carry male flowers on one plant and females on another are said to be dioecious, and the flowers unisexual. Those which carry separate male and female flowers on the same plant are described as monoecious; and those which possess flowers in which both male and female parts are present are known as hermaphrodites, and the flowers are described as bisexual.

Certain plants such as Japanese cherries, Jews mallow (*Kerria japonica*) and the mock orange (*Philadelphus*) include types with double flowers. These are not capable of producing a seed because their extra petals have replaced the stamens and the carpels which are the reproductive organs. Seed, however, is produced to a limited extent in some double flowers such as chrysanthemums and dahlias.

An interesting example of seed production in relation to doubling in flowers is observed in the ten-week stock which, being an annual, is always raised from seed. In certain strains a proportion of the flowers assumes a double form which is far more decorative. But double stocks do not produce seed, so it is essential to have some with single flowers for seed production. It is found that with certains strains of these stocks, double and single forms may be recognized when they are seedlings and separated. The doubles are usually a lighter green than the singles and show up best in cool temperatures. This allows the singles to be discarded or planted in a nursery bed for seed production, while the double-form seedlings are set out in beds or borders for floral effect.

Vegetative increase

Vegetative methods simply involve isolating portions of a plant, whether of the stem, root or leaves, and inducing them to grow and develop into separate individuals. In this case the resulting plants are in a very real sense 'chips off the old block' and are not distinct original specimens as are seedlings. Because no sexual activity is involved, vegetative propagation is termed asexual.

The fact that the new plants are part of the old and consequently identical to it ensures absolute uniformity and conformity to type. This is the principal advantage of vegetative propagation as compared with seed, and is a factor of extreme importance in horticultural practice. It has made possible the standardization of many valuable garden varieties. Indeed, the majority of these began their existence as single individuals. Thus the many thousands of 'Worcester Pearmain' apple trees have all been derived from an original seedling by vegetative propagation and together constitute what is called a 'clone' (a group of identical plants reproduced vegetatively from one parent).

If it were not possible to increase apples vegetatively each new variety would have to be 'fixed' before it could be relied upon to 'breed true' from seed. That is to say, it would have to be crossed with similar plants, the best ones selected and re-crossed or self-pollinated until eventually one individual would produce consistent offspring. Such a process, if at all possible, would take many years of hybridizing and selection. This is true not only of apples but of all our modern varieties of fruit. It applies also to the innumerable hybrids of flowering plants which are found in every garden. Few of these could be raised true to type from seed, but immediately a new variety is produced vegetative propagation ensures its multiplication and the faithful reproduction of all its characters almost indefinitely.

Another advantage of vegetative propagation is that many plants so raised start off with an abundant food supply, and rapidly develop into mature and productive plants. Bulbs, tubers and corms, for instance, are comparatively large structures and may give rise to adult flowering plants within a few months of being planted. Such plants do not require the careful nursing in the early stages which is so essential with seedlings; further, they can compete on better terms with weeds and various plant pests. Even plants raised from cuttings or grafts usually reach maturity in a shorter period than do seedlings of the same species. Thus an apple, pear or plum when grafted comes into fruiting two or three years afterwards.

The principal disadvantage of vegetative propagation is that there is far more likelihood of disease being transmitted from the 'parent' plant to the new individuals than is the case with seed. This applies in particular to virus diseases which usually cause a gradual deterioration of the plant. Thus strawberry plants affected with certain virus diseases gradually become reduced in size and less productive until they are practically worthless. Runners taken from diseased plants are, of course, also affected and so the disease is retained and increased by propagation of this type.

Spores

There is one method of propagation which is quite distinct from either seed or vegetative means. This is the increase of plants by what are called spores. These are minute reproductive units somewhat similar to seeds. They are in fact the 'seeds' of flowerless plants, of which mosses and ferns provide examples, as well as fungi such as mushrooms. Unfortunately, spores are responsible for the spread of many serious fungus diseases that affect plants.

The gardener uses spores only for the increase of certain ferns. In this case the spores are found on the underside of the fern leaves which are more correctly called fronds. Ripening is indicated by the sori (spore clusters) turning brown. It is usual to collect a few fronds and enclose them in a bag. In a day or two it will be found that a plentiful supply of spores has been released.

Notes on fern propagation will be found in Chapter 19.

2
Facilities and Equipment for Propagation

The use of the word nursery for a garden specializing in the propagation of plants indicates that plants in common with children require special care and attention in the early stages of their development. Given these conditions results are more certain and there is less delay in the production of healthy individuals. The would-be propagator should, therefore, provide himself with the best possible facilities for his task.

A great many plants, whether increased from seed or cuttings, may be raised in the open. Obviously, a plot situated in a warm sheltered position is most suitable for this purpose. A sunny corner of a walled-in garden is ideal, but any plot facing south or south-west and protected from cold winds may be equally good.

Ideally, the soil should be light and free working, for if the land is heavy it is difficult to secure that loose, friable surface usually described as a good tilth, which is so essential for seed sowing. Good drainage is the first principle of soil management. Wet land is difficult to handle, slow to warm up in the spring and, consequently, does not favour early germination of seed or quick growth. Good cultivation and proper manuring do make some amends for soil defects, and it is important to keep the plot free from weeds.

A good water supply in the vicinity of a plot used for propagation is an asset, and some water is essential if there is any glass cover. Water from the mains or from a deep well is preferable to a shallow well, pond or river as such sources of supply may carry disease infection, for example, the damping-off disease of seedlings.

In relation to the site some thought should be given to the prevention of the various pests of seeds and seedlings. Birds will devour both and are likely to be numerous where there are a lot of trees. Rabbits are also a menace to many young plants and, if present, should be excluded from the plot with wire netting fences whose bases are sunk into the ground for a depth of 45 cm (18 in). Rats and mice are extremely injurious to seeds and if they are observed steps must be taken for their destruction.

In private gardens it is not always possible to have a special plot devoted solely or mainly to propagation, but if this can be arranged it is a great advantage. Such a place is usually called the reserve garden, and while being convenient for general propagation also enables the gardener to maintain a reserve of plants for various purposes, such as filling in the gaps in a herbaceous border or rock garden. Whether or not one is fortunate enough to have a reserve, always select the choicest spot for any propagation work undertaken. This is particularly important for raising seedlings in the early spring.

The advantages of a greenhouse
A greenhouse, especially if heated, greatly extends the variety and scope of propagation. All kinds of structures are used for this purpose, but the most suitable type is one that is strong, durable and well designed, and which admits the maximum amount of light. This is important, particularly where most of the propagation is done in the very often dull short days of late winter and early spring. Good light conditions are essential for the sturdy growth of seedlings and as

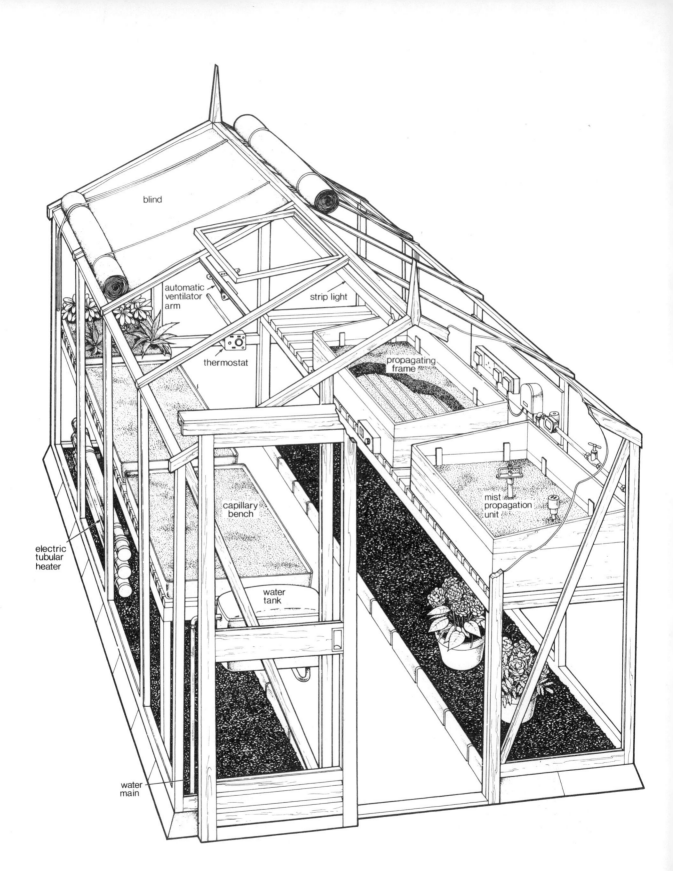

blind

automatic ventilator arm

strip light

thermostat

propagating frame

capillary bench

electric tubular heater

water tank

mist propagation unit

water main

much as possible of the limited light available should be allowed to enter the greenhouse.

Before erecting a new greenhouse there is an opportunity to choose a design which will allow maximum light admittance. Panes should be large, preferably not less than 60 cm (24 in) wide. The framework supports should be as slender as the necessity for a certain degree of strength and stability in the structure will allow. Such materials as aluminium alloy and western red cedar fulfil these conditions best.

Experiments have shown that a greenhouse which runs from east to west is best for light admission in winter and early spring, and consequently preferable for raising seedlings at this time of the year. Moreover, it is important that the propagating house should stand well away from any kind of shade such as a tree or buildings, including another greenhouse. Some shelter at a distance, however, from north and north-east winds is desirable.

Of course, one has often to use a greenhouse already in being and in such circumstances everything possible should be done to minimize light obstruction. It goes without saying that the glass should be kept scrupulously clean both inside and out during the propagation period.

Staging can be constructed with a top surface of welded steel wire mesh, preferably galvanized, supported on a wooden or steel framework. The steel mesh is usually 8 cm by 2.5 cm (3 in by 1 in) and is made from 10 or 20 swg wire.

The amateur will rarely have a greenhouse used solely for propagation. In the spring most or all of the house may be used for this purpose but at other times only a part of it is required. An ideal greenhouse in which to propagate plants is shown in Fig. 5.

Greenhouse heating

An unheated greenhouse often serves a useful purpose in propagation but its value is greatly increased when artificial heat is added. The following are the most effective ways of heating a greenhouse:

Boilers A system of water-filled heating pipes connected to an oil, solid-fuel or gas boiler is the most expensive system to install. The pipes are usually situated under the staging and the boiler is positioned outside the greenhouse in some form of cover which provides shelter from the weather. This type of heating system is extremely good in large greenhouses but is slow to respond in smaller areas.

Natural gas heaters Natural gas or propane burners have become popular in recent years (Fig. 6a). They can be thermostatically controlled and are relatively cheap to install and to run. These heaters give off both carbon dioxide and water and a little ventilation should be given at all times to allow combustion to take place. A professional fitter should be called in to connect up the heater, and in the case of natural gas, a pipeline will have to be laid to the greenhouse. Recent tests have shown that these flucless heaters do give off nitrogen oxide and ethylene, both of which are damaging to plant growth. In many instances no difference in crop development will be noticed, but there is a chance that growth could be impaired, particularly if ventilation is poor.

Electric heaters Even if they are not the cheapest to run these heaters are certainly the cleanest and most easily controlled. Fan (Fig. 6b) and tubular types are the most common. The first should be positioned where it will blow out its warm air most effectively; the tubular heaters are usually positioned in banks underneath the staging. Both types are best fitted with thermostats to ensure efficiency and economy, and they will produce a dry heat which is especially beneficial in winter when moisture-loving fungus diseases can be troublesome.

Bottom heat for propagating beds in greenhouses or frames is a great advantage both for rooting cuttings and for seed propagation. If there are hot water pipes beneath the staging this may be adequate, but to provide additional heat or in an unheated greenhouse or frame, electricity is the ideal method.

Insulated soil-warming cables fulfil this purpose and may be connected to the mains without a transformer. To heat the compost or propagating medium the cables are laid on about 4 cm (1½ in) of sand made firm and level. They are then covered with a similar depth of sand. A 10 cm (4

Fig. 5. This greenhouse is well equipped for the amateur propagator, possessing both mist and ordinary propagating frames, automatic watering and ventilation, plus heating apparatus. Shading is also vital to prevent dehydration of the young plants in bright spells of sunshine.

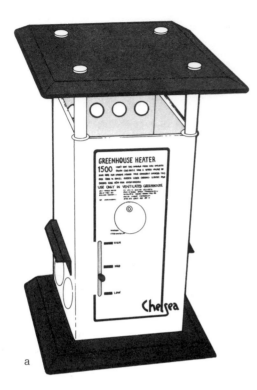

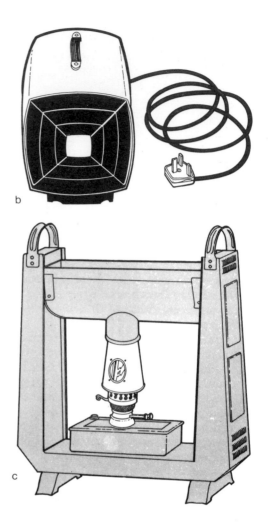

Fig. 6. Greenhouse heaters. (*a*) Thermostatically controlled heater available in natural gas or propane versions. This model, with an output equivalent to 1.5 kW, is suitable for greenhouses up to 9.3 sq m (100 sq ft). (*b*) Electric fan heater – clean and easily controlled. (*c*) Paraffin heater – simplest to instal.

in) deep layer of the propagating medium is placed on this or, alternatively, boxes or pots are filled and stood on the sand. It is a good idea to fill spaces between pots or boxes with moist peat to conserve soil warmth. This type of heating should be thermostatically controlled (by a rod thermostat) with the object of maintaining 18 to 21°C (65 to 70°F) in the compost. To achieve this a loading of 10 to 12 watts per 30 cm sq (per 1 sq ft) is required.

Heating cables may be used to heat a section of the greenhouse bench which can be covered with a polythene shroud. Overwintering plants can be placed on the protected bench and kept alive and warm through the winter at little cost.

Paraffin heaters Their cheapness and portability has made paraffin heaters (Fig. 6c) very popular in the amateur's greenhouse, but they create excessive humidity and give off damaging fumes and for this reason can only be recommended where no other type of heating system is practicable. Hollow or water-filled extension pipes can be attached to these heaters to aid heat distribution, but ventilation must be applied at all times if foliage scorching is not to become a problem. It is particularly important with this kind of heater that wick cleaning is practised regularly and that only the best quality paraffin is burned. Keep the heater level and out of draughts.

Using frames
Frames enable the propagator to provide close humid conditions both inside a greenhouse and in the open. They are also much cheaper to heat

than the larger area of a greenhouse. Some greenhouses used for propagation are provided with permanent frames, but it is usually more convenient for the amateur to construct a frame to stand on the staging. This may be made from boards 30 cm (12 in) high at the back and 23 cm (9 in) high in front. A glazed 'light' or a polythene shroud provides a cover.

A similar type of frame may be erected in the garden, preferably in a sheltered shady position such as against the sunless side of a north wall. Sometimes frames in the open are provided with double cover consisting of a sheet of polythene over which one or more lights, according to the frame length, are placed. Dutch lights (large glazed frames measuring approximately 1.5 m by 75 cm (5 ft by 2½ ft)) are excellent for this purpose.

Cold frames in the open have many uses in propagation. Softwood cuttings from herbaceous plants and shrubs can be raised in them during the warmer months; in the autumn semi-mature or hardwood cuttings of plants such as conifers may be planted in them. Frames are also useful for starting early vegetable crops and other plants from seed. Sowing can also be done under the continuous type of cloche to give seedlings an early start.

Plastic tunnels

A simple but effective method of protecting cuttings was developed by the Glasshouse Crops Research Institute at Littlehampton. This consists of a plastic tunnel similar to tunnels made for strawberry protection. The 1 m (3 ft) wide wire hoops which support the polythene are made from 8 swg wire and are erected at 75 cm (30 in) intervals. A white translucent 250 gauge polythene sheet 1.75 m (6 ft) wide is then stretched over the hoops and is held in position by fastening two lines of polypropylene bailer twine to alternate hoops on either side of the tunnel (Fig. 7). The technique of striking cuttings under polythene is described in Chapter 11.

Propagating units

There is now a wide range of self-contained propagators, heated and unheated, on the market. The simplest type consists of nothing more than a standard plastic seed tray with a transparent plastic cover known as a propagator top, which is made to fit over the tray. One firm provides this type in three sizes, the smallest one being 23 cm (9 in) long by 15 cm (6 in) wide and nearly 18 cm (7 in) high overall. The largest one is nearly 60 cm (2 ft) long by 30 cm (1 ft) wide and just over 23 cm (9 in) high. These have propagator tops made from clear polystyrene and are fitted with one or two adjustable ventilators.

Larger and more sophisticated propagators are available. One of these is rather like a large glass case with a strong metal framework. It is nearly 1 m (3¼ ft) long, about 60 cm (2 ft) wide and 60 cm (2 ft) high. This propagator is fitted with sliding glass panels both front and rear making for easy access. Similar models are available in small sizes and each has a damp-proofed base about 15 cm (6 in) deep.

Most propagators are provided with heating equipment, but in some cases this is optional. If bottom heat is installed a sandbed can be arranged in the base as previously described. Some propagators are also fitted with air heating by means of warming cables. Both air and bottom heat are usually thermostatically controlled and

Fig. 7. Polythene tunnel cloches: inexpensive, afford protection for seedlings from frost in early spring and are ideal for getting many vegetables off to an early start.

this is always desirable. Even a single tray with or without a cover can be heated from below. This is done by standing the tray on a proprietary electric heating panel.

Electrical heating of propagators presents no problems for the amateur so long as there is an electric socket somewhere convenient. Propagators are usually stood on the greenhouse staging. A small propagator can be positioned in a room window.

The amateur can easily construct a simple type of propagator for himself. Make a sandbed in the base of a 23 cm (9 in) deep box with bottom heat and provide a layer of about 8 cm (3 in) of rooting compost. Over the bed erect a light framework to carry a plastic sheet. A deeper box may have a sheet of glass laid over it as a cover.

Growing rooms

A growing room is an insulated building from which natural light is usually excluded. Illumination is provided by artificial means. Growing rooms are now widely used commercially for the producton of seedlings such as bedding plants, tomatoes and cucumbers. The seedlings are usually grown in trays or pots stood on benches. To save space the benches are usually installed in tiers being vertically about 75 cm (30 in) apart. They are usually 2.4 m (8 ft) long by 1.1 m ($3\frac{1}{2}$ ft) wide.

Each bench is illuminated with 2.4 m (8 ft) long 125 watt fluorescent tubes mounted 45 cm (18 in) above the bench. Seven tubes over each bench provide a light intensity of 500 lumens per 30 cm sq (per sq ft) which is adequate for bedding plants. Tomatoes and lettuce require 1,000 lumens per 30 cm sq (per sq ft) which is given by thirteen tubes to each bench. Bedding plants grow best with continuous illumination but tomatoes and lettuces should be restricted to 12 to 16 hours per day. The heat from the fluorescent tubes usually maintains a temperature of at least 21°C (70°F) and excessive temperatures are prevented by the use of fans installed in the building.

Because both light and heat can be standardized and maintained uniformly over a given period this method allows a high degree of control over seedling production and is very successful.

The amateur who wishes to try raising seedlings by this method will notice that light of a high intensity is required. A bench in a reasonably insulated shed could be used. Fluorescent tubes

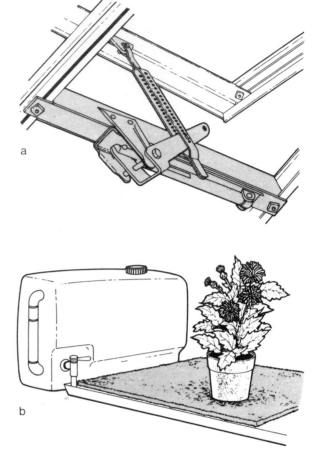

Fig. 8. Automation in the greenhouse. (*a*) This type of automatic ventilator arm needs no external power supply, being activated by the expansion and contraction of a heat-sensitive material. (*b*)The capillary bench watering system, which ensures that the surface on which the plants stand is kept permanently moist.

of the type mentioned 1.2 m (4 ft) long instead of the 2.4 m (8 ft) size may be more convenient but must be concentrated as mentioned above. Supplementary lighting can also be fitted in greenhouses to provide continuous overnight lighting for bedding plants and cucumbers and to lengthen the natural day for tomatoes and lettuces.

The automatic greenhouse

Today, the modern greenhouse can be almost completely automated, thus assisting propagation. For instance, by the use of thermostats air and bed temperatures can be maintained at any given reading according to requirements. Similarly, automatic ventilation allows the ventilators to open and close in relation to temperature (Fig. 8a).

Two main systems of watering can be largely automated. The first is the capillary bench which may consist of one or more capillary trays. These are filled with sand or nylon fibre matting, and water is fed in at the bottom edge. The water level is maintained by a valve at the side of the bench, the valve being kept supplied with water from an overhead tank (Fig. 8b). Potted seedlings are stood on the moist sand or matting from which water is drawn up by capillarity. For this purpose plastic pots are best for they have many holes in their bases and the compost will be in firm contact with the moist medium. If clay pots are used they must be fitted with fibreglass or nylon fibre wicks instead of crocks.

A second method of watering consists of drip or trickle irrigation whereby water is supplied to each pot or plant by individual nozzles or tubes. This system (sometimes known as 'spaghetti tube' watering) can be controlled from the mains by turning the tap on or off as required, or by means of a time-switch. The plants are, however, all watered at the same time and there is always a risk of over- or under-watering.

The potting shed and other facilities

A potting shed is regarded as essential in a nursery and is a great asset in any garden, particularly where composts and fertilizers are made up rather than bought ready mixed. Gardeners who raise most of their own plants in quantity will find such a building of great use. A well-lit building is most suitable and if it can be heated so much the better. The potting bench should be smooth and firm and slightly above waist height. It is an advantage to have a level concrete floor for mixing composts. Inside the shed, or adjacent to it, ample storage space should be provided for the various materials used in propagation. These include loam, peat, sand, fertilizers and chalk or lime. Sterilized soil should be stored separately in a closed bin to prevent contamination.

Some means of sterilizing soil for propagation is a valuable facility. Electric soil sterilizers are very convenient for treating small quantities. One type on the market holds about a bushel of soil (the amount contained in a box 25 cm by 25 cm by 55 cm (10 in by 10 in by 22 in) which the makers claim is sterilized in about 1½ hours. These sterilizers are not expensive and appear to be most suitable for the amateur. For further details see pp. 52 and 53.

Pots and trays

The principal receptacles used in propagation are various sizes of wooden or plastic trays. The standard size plastic tray is 35 cm long by 23 cm wide and 6 or 8 cm deep (14 in long by 9 in wide and 2½ or 3 in deep). It may be used for seed-raising or cuttings. Various other types of boxes are quite suitable but it is some advantage if all are of standard size as this facilitates their arrangement on the greenhouse staging. Wooden trays may be obtained from greengrocers and fishmongers. They are cheaper than their plastic counterparts but less durable and more difficult to keep clean. Consequently they are less hygienic. Plastic pots are available in a wide range of sizes and are generally preferable to clay pots being cleaner, lighter and more durable. The size specified represents the diameter across the top. The 8 cm (3 in) and 11.5 cm (4½ in) size pots are generally most suitable for propagation. Half-pots or pans (which are just as wide as pots but only half as deep) are available in equivalent sizes. Trays are generally used to raise moderate or large numbers of seedlings or cuttings, such as annuals and chrysanthemums. Pots and pans are ideal for small quantities of either cuttings or seed.

Tools

Equipment used for various operations in propagation includes most of the ordinary garden implements, such as a spade, fork, rake and hoe. A trowel is very necessary for planting, and one of the most important tools is the dibber. Small dibbers for glasshouse work vary in size from those of pencil thickness up to 1 cm (½ in) in diameter. Large dibbers for use outside are made from 2.5 to 3 cm (1 to 1½ in) in diameter dowelling or from old spade or fork handles. Dibbers should be made of hardwood.

A good knife is the propagator's best friend. There are several types, but so long as it is fairly strong, has a folding blade that takes a keen edge, and a handle that provides a good grip, it should be suitable. A special knife is used for budding.

Secateurs are often handy to supplement the knife. Sieves of various sizes are necessary for seed-sowing.

Patters or firmers are very useful for levelling and firming soil in receptacles and may be round, square or rectangular in shape. A pot bottom serves for firming the soil in pots. A watering-can provided with a coarse and fine sprinkler head or 'rose' is essential, and a hand sprayer with an adjustable nozzle is very serviceable. Last and no less necessary than the various tools and implements is a good supply of labels. Plastic labels have replaced wooden ones, but those made of anodized aluminium are perhaps the most durable and, like plastic ones, can be scrubbed clean and re-used. The 10 cm (4 in) size is convenient for indoor use, but in the garden a minimum length of 23 cm (9 in) is desirable.

3
Hybridization and Plant Inheritance

Throughout the ages, plants have been slowly improved by gardeners who have constantly selected the finest specimens for seed production. No doubt artificial crossing or hybridization as a means of promoting improvement has been used since time immemorial, but up to the beginning of this century nothing at all was known about the principles involved which could serve as a guide to any would-be plant breeder.

Today, we know that plant characters are passed on from generation to generation in accordance with a definite natural law which is called Mendel's Law, in honour of its discoverer, G. J. Mendel, (1822–84). Although Mendel gave some publicity to his discovery in 1860, its importance, unfortunately was not realized until the year 1900.

Mendel's field of operation was a monastery garden, and he used the edible pea for his researches. This plant normally sets its seed by self-pollination and is, therefore, particularly suitable for plant breeding experiments. Mendel crossed a tall pea plant with a dwarf type and in the first generation of seeds obtained from this cross (usually described as the F_1 generation), all the plants were tall. When the F_1 generation grew to maturity and flowered it was self-pollinated (selfed) and the resulting progeny (termed the F_2 generation) revealed a variation of approximately three tall plants to one dwarf. All the dwarf peas, when selfed, were found to breed true to type as did one in three of the tall peas. The remainder—half the total—gave results exactly similar to all the F_1 generation.

Mendel's explanation of this result was that, corresponding to each inherited character, such as tall stems in peas, there is a factor (now called a 'gene') which is transmitted from the parents to their offspring via the germ cells, that is, the pollen and the ovules. As every individual arises as the result of the fusion of male and female germ cells, or 'gametes', it will receive a gene from each in relation to the particular character. Consequently, when a plant breeds true for a given character it may be assumed to have had a double dose of the appropriate factor.

A pure breeding tall pea plant is an example and may be represented by the letters TT. Similarly, a dwarf pea plant which reproduces itself true to type may be designated tt. The use of the same letter as a capital and in its small form is to illustrate the relationship existing between these two alternative or contrasting characters, namely tallness and dwarfness. Mendel termed constant differentiating characters of this type a 'pair of allelomorphs'. True breeding plants like those mentioned are said to be 'homozygous' for the character in question.

Mendel postulated that each gamete carries one unit or gene only relative to the particular character; consequently a gamete from the tall peas may be indicated by the single letter T and all those of the dwarf plants as t. Obviously, a cross between the tall and dwarf varieties would result in individuals having both tall and dwarf factors in their hereditary constitution, that is, a single dose of each. Such plants are described as 'heterozygous', and may be designated Tt to indicate their mixed origin. Naturally, as we have already seen, such plants could not breed true for either of the characters concerned as half their gametes would be T and half t. In fact self-pollination of heterozygous plants results in a sorting out or segregation of the genes, giving in our example three possible combinations, namely TT, Tt and tt in the ratio of 1:2:1. It will be noted that all the plants in the F_1 generation appeared tall although they had inherited both tall and dwarf genes in equal proportion. Mendel therefore called tallness in peas 'dominant' and dwarfness 'recessive' characters. This quality also affected the F_2 generation in that 75% of the plants had tall stems, although only 25% of the total number would breed true for tallness. Obviously, all the dwarf plants would breed true for dwarfness, as the fact that they showed this quality at all indicated that they had had a double dose of the appropriate gene. This rule will, of course, apply to any recessive character.

Mendel's experiments with peas revealed that these plants possess several other pairs of allelomorphs, including white or coloured flowers; green or yellow pods; and round or wrinkled seeds. In each case the character mentioned first is the dominant one.

This conception of dominance, while being very important, is not by any means always applicable. Thus certain characters are dominant over some but not over other characters. Again, two genes carrying different characters may affect the offspring in such a way as to give an intermediate result. For instance, when red antirrhinums are crossed with a white variety all the progeny in the F_1 generation are pink. If the pink flowers are selfed the genes segregate to produce red, pink and white flowered plants in the ratio of 1:2:1. Another example is a cross between *Primula sinensis* which has large wavy flowers and its variety *stellata* with small star-shaped flowers. This cross results in blooms intermediate in size and shape, and the hybrid when selfed behaves in simple Mendelian fashion like the pink antirrhinums.

When two pairs of allelomorphs are involved the results are naturally more complicated. For example, if we cross a tall pea plant with round seed and a dwarf plant having wrinkled seed, all the progeny will have tall stems and round seed (Fig. 9a). From these F_1 hybrids the male and female gametes carry genes in equal numbers which may be indicated as TR, Tr, tR and tr. When the F_1 generation is selfed the possible

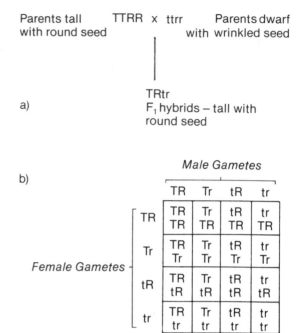

Fig. 9. Mendel's Law exemplified by crossing tall pea plants having round seed with dwarf plants having wrinkled seed. (*a*) F_1 generation (*b*) F_2 generation.

combinations are shown in Fig. 9b.

Fig. 9b shows the following results applicable to every 16 plants:

9 tall plants with round seed—one will breed true.

3 tall plants with wrinkled seed—one will breed true.

3 dwarf plants with round seed—one will breed true.

1 dwarf plant with wrinkled seed which will breed true.

This result shows that two new combinations of characters have appeared, namely: tall plants with wrinkled seed and dwarf plants having round seed. In each case one out of three such plants will breed true for both qualities. It should be pointed out here that, except when dealing with recessive characters, the plant-breeder cannot tell from the appearance of the plants which will breed true. This can only be ascertained by selfing again and raising a third generation. In fact, it may be necessary to continue in this way for several generations, always selecting those that have the required characters. Great advances in the science of genetics have been made since Mendel's work was first generally appreciated, and the above account is a rather simplified picture of inheritance in the plant world. Thus, every plant character is usually controlled not just by one gene but by a number of genes; moreover, one gene may affect several different characters. Mendel's Law, however, is still regarded as the basis of genetics and is of immense importance to the plant breeder.

Mendel's success was due to the fact that he studied plant characters separately, carefully noting how each was distributed among succeeding generations. In plant improvement this underlines the importance of having a definite objective in view and of concentrating on a particular character at a time. To attempt more than this at once is likely to result in complete failure.

Another lesson arising from Mendel's Law is the importance of the F_1 generation, and indeed of later generations. Apparently that is often overlooked both by amateur and professional hybridizers, and applies particularly to such popular subjects as chrysanthemums, dahlias and roses which lend themselves to quick and easy methods of vegetative propagation.

In many nurseries it appears that the usual practice with plants of this type is to cross two varieties and select what are considered to be improved sorts from the first generation. These are henceforth increased vegetatively and the remaining seedlings discarded. So it is that the nurseryman in quest of new varieties is for ever crossing and re-crossing older varieties. Perhaps he would get more interesting results by selfing the F_1 generation and in some circumstances continuing in this way for several generations. Careful records should be kept of characters appearing in each generation.

It should also be remembered that most varieties of the subjects mentioned are the results of repeated crossing. Consequently, if one selects and selfs a variety of, say, roses, the chance of getting a good variety is probably just as great as by crossing that variety with another.

Hybridization is, of course, essential when one wishes to combine in one individual or strain distinct qualities possessed by two or more individuals or strains. Therefore, the most striking results are likely to arise when two fairly distinct varieties are crossed. Thus, it may sometimes be advantageous to secure new stocks of certain plants from abroad, with the object of introducing 'new blood' into a nursery collection.

As an alternative to selfing, back-crossing is a method advocated by geneticists. This means crossing back the F_1 hybrids or later generations to the original strain or variety one is attempting to improve. For instance, if a standard variety of tomato was crossed with a wild type, perhaps with the object of conferring upon the former immunity from disease, very likely the F_1 generation would consist of worthless hybrids. Adequate improvement might, however, be effected in time by back-crossing for several generations with the original superior parent. Ultimately perhaps a type would be secured capable of producing a high yield of good quality fruit and which had also acquired from the wild type immunity from disease.

Another consideration in plant breeding is whether the subject normally sets seed by self- or cross-pollination. Those in the latter category, such as brassicas and beetroot, have usually a very variable hereditary make-up. Here it is advisable first to purify the strains it is intended to cross by raising several successive generations of each from selfed plants. This is found usually to bring about a loss of vigour, but when uniform offspring are produced consistently such plants are termed a pure line. Two pure lines are often crossed to combine their good qualities in one plant or strain. Crossing in this case usually results also in a pronounced increase in vigour.

When dealing with plants which usually produce seed by self-pollination, such as peas, French beans and tomatoes, strains of these often consist of a number of pure lines which have arisen naturally. Improvement of such plants should, therefore, in the first place be done mainly by selection. Crossing of the best types can then be attempted with the usual aims in view.

Mutations or 'sports'

Another manner in which new varieties sometimes arise in plants is by what is called 'mutation' or 'sporting'. Thus a plant with white flowers may suddenly produce a branch having pink flowers. Sometimes the seed from such flowers will breed true and we have a new variety. More often the new type will not come true from seed, but can be increased vegetatively. Sporting often occurs in chrysanthemums.

Plant improvement effected by mutation and by natural or artificial hybridization is the result of a change in the plant's hereditary constitution, and consequently may be passed on from generation to generation. Any improvement, however, which is due to environmental factors, such as soil or temperature, cannot be so transmitted. Sometimes it is not easy to distinguish between the effects of inheritance and environment, but the difficulty can, to some extent, be minimized by growing plants intended for breeding under conditions which are kept as uniform as possible.

F_1 hybrids

A casual glance through almost any seed catalogue indicates that a great many varieties of both vegetables and flowers are described as F_1 hybrids. This simply means that the seed has been derived from crossing two distinct parents and will give rise to a first generation cross. Again the object is to combine in one individual certain desirable characters contributed by each parent. This will always occur provided the required characters are dominant, and the process is called heterosis.

When such a cross has been effected with good results it can be repeated if and when plants of the original parents are available. In this way a good variety is obtained without further experimental work on hybridization. The disadvantages are that the breeder must ensure a supply of either seed or plants of both parents true to type. Also the production of F_1 hybrids often involves careful hand pollination and isolation of the plants used for breeding. This is why the seed of F_1 hybrids is usually much more expensive than ordinary varieties.

The obvious popularity of F_1 hybrids indicates their merits. Thus the combination of good qualities from each parent has often resulted in some outstanding novelties in both flowers and vegetables raised from seed. F_1 hybrids are often more vigorous and healthy than either parent and therefore easier to grow. Disease resistance is a common character.

The outstanding characteristic, however, displayed by F_1 hybrids is their extraordinary uniformity in habit of growth and in constitution. This means that one can plant F_1 hybrids with greater confidence in the knowledge that the plot or bed will not be marred by weak or untypical plants.

The number of F_1 hybrids available is increasing year by year. In vegetable crops good examples will be found in Brussels sprouts, summer and autumn cabbage, tomatoes, cucumbers, melons and vegetable marrows. In flower crops quite outstanding F_1 hybrids are included in antirrhinums, begonias (fibrous rooted), African marigolds and petunias.

Plant breeding techniques

The plant breeder must have some knowledge of the structure of flowers and particularly of the reproductive organs (Fig. 10). The 'stamens' are the male sex organs and are found inside the 'perianth' (a collective term used for the sepals and petals). Each stamen has a swollen tip—the 'anther'—which produces the 'pollen' in which the male gametes are contained.

Fig. 10. Reproductive organs of a typical flower, in this case a fuchsia.

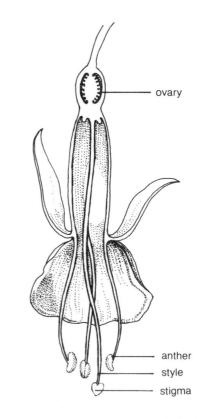

ovary

anther
style
stigma

Fig. 11. Before and immediately after cross-pollination, the flower of the female parent should be encapsulated in a translucent paper bag to prevent the entry of unwanted pollen.

stigma of a flower is pollinated by pollen from the same plant, this is called 'self-pollination', but if by pollen from a different individual it is called 'cross-pollination'. Plant breeding usually involves artificial pollination and also the taking of precautions to ensure that pollen from only the selected male parent reaches the stigma of the chosen female parent.

The usual procedure in plant hybridization is briefly as follows:

1. Select the flower to be used as the female or seed parent and, usually before it has opened, carefully remove the stamens. This emasculation will prevent self-pollination and timeliness is essential. The operation is not, of course, necessary with unisexual flowers.

2. Enclose the flowers of the seed parent and also of the selected male parent each in a paper or muslin bag to prevent foreign pollen coming in contact with either (Fig. 11).

3. When the stigma of the female parent is receptive, that is, slightly moist and sticky, transfer pollen to it from the male parent. This may be done with a clean camel hair brush, or the anthers may be brought into direct contact with the stigma if the male flower is broken off.

4. Again enclose the pollinated flower and keep it covered until the ovary is seen to be enlarging. This indicates that fertilization has occurred.

5. Save the seed and sow when appropriate.

The 'carpels' as they are known singly, collectively termed the 'pistil', occupy the centre of the flower. In some plants such as peas and beans the pistil consists of one carpel, i.e. the pod, but in other plants like buttercups there are many carpels. Each carpel is composed of the 'stigma', the 'style' and the 'ovary'. The last is the female organ containing the 'ovule' or 'ovules', which may, if fertilized, become seed.

Most flowers possess both stamens and pistil and are termed 'bisexual'. A few flowers, however, such as those of the vegetable marrow, have the male and female organs borne in separate flowers which are called 'unisexual'.

Pollination is the transference of pollen, naturally or artificially, from the anthers to the stigma, and normally results in fertilization. When the

4
Harvesting and Storage of Seed

Can the home gardener save seeds from his own garden and get good results from them? Certainly this is done successfully in many cases, but there are complications, and with some kinds of plants it is definitely inadvisable. The chief difficulties are as follows. Firstly, in small gardens there is insufficient room to allow any space to be set aside especially for seed production. Secondly, in many districts it is impossible to secure proper ripening and drying owing to moist weather in the autumn. Thirdly, there is always the risk of indiscriminate cross-pollination giving rise to worthless hybrids.

If one attempted to save seeds from cabbages and Brussels sprouts growing side-by-side, very likely some or all of the seedlings would be a cross between these two vegetables. All the brassicas are prone to crossing in this way, and even if only one type is grown there is always a risk of cross-pollination from a neighbour's garden or elsewhere. Moreover, the varieties of most vegetable plants set seed readily with other varieties of the same kind, and in a private garden it is rarely possible to provide adequate isolation.

Vegetables
Seed saving of vegetables in private gardens should usually be limited to those vegetables that normally set seeds with their own pollen, in other words, they are self-fertile. Even then there is a risk of some cross-pollination unless one variety only, of each of the vegetables attempted, is grown. It is also advisable to restrict home seed-saving to vegetables that ripen their seed early and are relatively easy to harvest; such crops include peas, broad beans, runner beans, French beans, lettuces and tomatoes.

In saving seed from vegetables, as far as is possible select only those plants that are healthy, true to type and conform to the general standard of the variety. It should be emphasized that any improvement in plants or in their productivity which is due to good cultivation is not passed on to their offspring. Everything possible should be done to prevent pollination of inferior types. If, for instance, one is saving seed from peas it is advisable to look along the row when the crop is in flower and remove plants that differ significantly from the typical variety. Tomato seed should be saved only from the best plants as regards yield and quality.

For some vegetables harvesting is not difficult. Peas and beans are good examples. However, if a good sample is to be secured, dry warm conditions are necessary at ripening time. Early sowings are, therefore, preferable for seed crops. As a rule it is best to set aside a row or part of a row for seed, but if only a small quantity is required pick the upper pods for cooking and reserve the lower ones for seed.

When the pods turn yellow and show signs of splitting they are ripe and may be picked off singly. Sometimes the plants are pulled first and hung up outside, or in a shed to dry. Broad beans are left until the pods turn dry and leathery, when they are pulled and dried.

In wet weather French and runner beans are particularly difficult to ripen. Allow them to remain until the leaves turn yellow and begin to fall. They may then be pulled and dried.

Biennial vegetables for seed production are usually planted out in the spring, but are given more space than when grown as food crops. Selected onion bulbs and first-class roots of carrots, beetroot, turnips and swedes which have been stored over winter and protected from frost and damp are normally treated in this way. Leeks and parsnips, which are both hardy, are simply transplanted in the spring, the opportunity being taken to select the very best plants for this purpose. Brassicas are not usually transplanted and remain where they are for seed production. It should be emphasized again that to avoid cross-pollination only one variety of each vegetable should be attempted and never more than one kind of brassica.

When the seed-heads are ripe they are cut and thoroughly dried, taking precautions against loss of seed through shedding. Seed cleaning on a small scale is not difficult. Peas and beans are simply hand-shelled, rejecting seeds that are discoloured, damaged, diseased or immature. Other crops may be hand-threshed and cleaned by sieving (Fig. 12a) and other simple devices.

Seed saving from fruits usually classified as vegetables, such as cucumbers, tomatoes and vegetable marrows, is more troublesome than ordinary vegetables. However, these can be attempted by the amateur. To ensure a good set of seed for cucumbers and marrows, careful hand-pollination should be done immediately the female flowers open.

Tomatoes selected for seed extraction should be sound, well shaped and taken from robust plants which have produced a high yield of good-quality fruit. The same rules apply to marrows and cucumbers. Before attempting to remove the seed ensure that the fruits are absolutely ripe. Ripeness is readily indicated in tomatoes. Ripe marrows usually have a hollow sound when tapped, but ripeness in cucumbers is more difficult to ascertain, at least for the inexperienced. The best general rule is to leave the

Fig. 12. Extracting seeds. (*a*) Large seeds can be separated from capsules and detritus by sieving. (*b*) Seeds from fleshy fruits (e.g. tomato) can be extracted by fermenting the fruit, then squashing the residue in a sieve (*i*), dropping the residual mass into a jar of warm water (*ii*) and decanting off any flesh that is floating (*iii*).

fruits on the plant as long as is necessary or practicable, and if there is still doubt about maturity store them in a warm room for a week or two.

Tomato fruits are pulped in a wooden or earthenware vessel with a rammer. To facilitate the separation of the seed from the pulp the mixture is placed in a container of warm water and left in a warm place for two or three days. The flesh will ferment and the seeds will be easily separated from any that remains. Tip the pulp into a wide-meshed sieve and wash the seed and the small pulp through into a jar. The seeds sink to the bottom and the pulp is drained off (Fig. 12b). Washing is repeated two or three times and a final washing is given with the seed in a fine sieve

Seed of vegetable marrows and cucumbers may be extracted in a similar manner. With marrows the pulp and seed are scraped out when the fruit is cut lengthwise, while cucumbers should have the thick skin removed. Afterwards a rammer is used on the pulp which is then soaked as recommended for tomatoes. After the seed has been separated from the pulp they are spread on a piece of cloth to dry.

Ornamental plants

Plants grown for their decorative merits offer the greatest scope for home seed saving. Many shrubs and alpine plants are true species and will breed true from seed. Most annual flowering plants seed readily, but there is some risk of cross-pollination if several varieties of each are grown. However, with many annuals there is no objection to mixed colours.

The innumerable hybrids of herbaceous plants cannot be reproduced true to type from seed even if self-pollinated. Excellent seedlings can, however, be raised from many of these plants in a variety of colours. Examples are dahlias, lupins and delphiniums.

The seeds of many choice shrubs and alpines are often difficult to procure through trade channels. Moreover, and particularly with alpines, it is frequently advantageous if the seed can be sown as soon as it is ripe. The seeds of lewisias and primulas, for example, deteriorate rapidly in storage. The raising of such plants from home-saved seed is a fascinating pursuit and a source of new interest to the keen plantsman.

Various alpines ripen their seed during the summer and autumn, and careful observation is necessary to avoid the loss of valuable seed. Some androsaces and the lewisias ripen their seed unexpectedly and it is then immediately released and scattered. A number of rock plants seed very freely. These include *Dianthus deltoides*, rock hypericums, silene, lychnis and penstemon. Many primulas also provide seed in abundance which germinates freely if sown when ripe. Several of the bog primulas, such as *P. helodoxa* and *P. florindae*, reproduce themselves true to type from seed. Others may show some variation in colour, but this does not detract from their beauty. In wet weather any dead petals attached to the seed pods of certain alpines, such as several gentians and cyananthus, should be removed. This helps to prevent decay of the seed.

Shrubs also require careful watching as the seed is often discharged on ripening. Brooms, laburnum and tree lupins are examples. Many shrubs such as berberis, cotoneasters and the snowberry produce their seeds inside berries. When extraction is done it is usually advisable to sow at once or store the seed in sand so that it does not dry out.

The raising of bulbous plants from seed is very interesting but often requires patience. A wide range can be propagated in this way, common examples being gladioli, tulips and lilies which usually seed freely. The seed pods or capsules are collected when they turn brown and show a tendency to split open. Late-flowering bulbous plants which may not ripen their seed properly in the open should have their stems cut off and be hung in a warm greenhouse to complete the ripening of the seed. Lilies require hand-pollination to ensure fertile seed.

Many herbaceous perennials seed readily and provide a cheap means of securing plants in quantity, although the results may be variable. Plants worth collecting seed from are delphiniums, lupins, sidalcea, platycodon, campanulas, hemerocallis, alstroemeria, anthericum, irises such as *I. sibirica*, oenothera, various poppies and verbascum. Coreopsis and gaillardia are normally raised from seed and are best treated as biennials.

The seed of various annuals, such as clarkia and larkspur, may be saved, but there is always a risk of cross-pollination and consequent deterioration of the stock. Sweet pea seed is often home-saved successfully. Quite good results may be secured with seed saved from various biennials (or plants so treated) such as sweet William, wallflowers and Canterbury bells.

There is little object in saving the seed of fruit except for some special purpose, such as attempting to raise a new variety. If seed is required, however, the same rule applies about the necessity of ensuring ripeness. When the pips and stones are extracted they are usually stored in sand so that they do not dry out, until required, or they may be sown at once. With soft fruit—gooseberries, currants, raspberries and strawberries—the seed has to be separated from the pulp by washing.

Seed drying and storage

With the exception of seed secured from certain succulent fruits, such as the stone fruits and certain berried shrubs, all seeds should be thoroughly dried. This may be accomplished by spreading the various kinds on sheets of paper laid in various shallow containers. A warm greenhouse is an excellent place for drying seed, but sunny windows or airy sheds will also serve. When dried each lot should be carefully cleaned. A fairly fine sieve is useful for this purpose and this, together with the simple method of gently blowing on the seed, gets rid of most waste material. Labelling is an important operation and should be done first when collecting the seed. Write the name of each variety on a packet, and leave it with the seed whose name it bears. When dried and cleaned, they can be packeted.

Unless there is a risk of spilling, the packets should be left unsealed to allow free aeration for the seed. However, if seed has to be transported careful sealing is essential. Fine seed should be put in a small packet which is then placed inside a larger one.

The manner of storing seed greatly affects its length of life. The ideal store should be cool, dry and be provided with gentle ventilaton. In a warm, moist room the germination capacity of most seed declines rapidly. Exceptions are seed of the stone fruits and some others which need moist conditions, but these are usually sown or stratified (see p. 39) when ripe. The seed of certain other succulent fruits, however, such as tomatoes, marrows and egg plants, require dry conditions in storage.

Under ordinary good storage conditions the seeds of the majority of garden plants have probably the highest germination capacity in their first season after harvesting. Second year seed often germinates quite well, but retention beyond that can rarely be advised. Apart from the effect of storage conditions, length of life in seed is variable. Lewisias show a decline in germination capacity after six months, and primula and parsnip seeds also deteriorate rapidly. Other seeds retain their viability for a number of years. Thus the seed of gourds, carnations, dahlias and wallflowers often germinate well when five years old, and balsam after eight years. Poppy and turnip seed are noted for their long life, while certain species of legumes (pea and bean family) retain their viability for fifty years or more.

Experiments have shown that practically all seeds may be kept alive for periods much in excess of normal by storing in temperatures below freezing point under controlled conditions.

Some seeds, although they appear quite ripe superficially, are not properly mature when harvested, and are then difficult to germinate. Freshly harvested lettuce seed often germinates poorly and slowly, but after being retained in storage for a period the rate of germination improves. Apparently, the period of storage allows some internal change to occur in the seed, called after-ripening, which promotes germination.

It is sometimes claimed that when certain seeds are kept for a number of years the resulting seedlings crop better. For instance, vegetable marrows from second-year seed are said to produce a heavier yield than do plants from fresh seed.

It should be emphasized that the seed of a great many alpines, trees, shrubs and bulbous plants is best sown as soon as it is ripe. In this case the problem of storage does not arise.

Seed quality

If the gardener saves his own seed he is responsible for the results. What steps can he take to ensure that purchased seed is of first-class quality? Seed appearance is some guide in this direction. Good seed is usually of uniform size and colour, plump and has a fresh sweet smell. Inferior seed may be small, have a shrivelled appearance and be different in colour from the type.

The two important considerations relating to quality in seed are purity and germination capacity. Impurities in seed may take three forms as follows:

1. Inert matter such as dirt and fragments of dried plant remains.

2. Seed of a different type to that which the package purports to contain.

3. Injurious weed seed.

The first two types may be regarded as useless matter for which the purchaser has to pay. The third type is not only useless but detrimental to a garden. Certain plant diseases may also be carried on seeds.

Impurities are usually calculated on a weight percentage basis. Thus, if 454 g (1 lb) of seeds contains 28 g (1 oz) impurities this equals $\frac{1}{16}$ or just over 6%. It has been estimated that a 1% chickweed seed impurity represents about 45,000 seeds of this weed per 454 g (per lb) of the sample, and illustrates that even a low percentage of certain weed seeds can be very serious. To ascertain pecentage purity take a sample at random. This may vary in weight from 14 g ($\frac{1}{2}$ oz) for small seeds up to 227 g (8 oz) for the largest seed, such as broad beans. For accurate results a chemical balance is necessary. The sample is examined almost grain by grain and the 'pure' seed and the impurities are separated. The 'pure' seed is weighed and the percentage purity can then be calculated. For example, if the original sample weighs 57 g (2 oz) and the pure seed 50 g (1$\frac{3}{4}$ oz) the percentage purity is:

(a) for metric quantities:

$$\frac{50 \times 100}{57} = 87.7\%$$

(b) for imperial quantities:

$$\frac{1\frac{3}{4} \times 100}{2} = 87.5\%$$

(The reason for the minor discrepancy is that the metric quantities cited are only approximate.) It is usual to identify and classify impurities into the types specified.

Germination capacity is usually defined as the proportion by numbers of the pure seeds that are capable of producing healthy seedlings. A germination test requires attention to detail and constant observation. The sample usually consists of 100 pure seeds representative of the package, but it is usual to take three or four samples. The seeds are placed on damp blotting paper or damp flannel laid in a plate or saucer and covered with similar material. Keep in a warm room. A temperature of about 21°C (70°F) is suitable for most seeds. The samples may be examined every second or third day, taking care to keep the material uniformly moist but not wet. Seeds that sprout are removed and recorded and the test continued for usually 10 to 20 days. The total number of seeds that have germinated out of 100 of course gives the percentage germination for a sample and an average can be made with reference to the other samples.

Seed that germinates quickly and uniformly usually gives the best results, but certain seeds, such as lettuce, when freshly harvested, despite being slow in germinating, may be of excellent quality.

Obviously, the value of any lot of seed is related to both purity and germination capacity. A consideration of both factors gives what is termed the 'real value' of the seed. This is calculated by multiplying the percentage purity by the percentage germination and dividing the result by 100. Thus, if a given sample has a purity of 80% and the germination capacity is 90%, the real value is:

$$\frac{80 \times 90}{100} = 72\%$$

45.25 kg (100 lb) of such seed contains 9.05 kg (20 lb) impurities and 3.62 kg (8 lb) of dead seed. If it is priced at say £2 per 454 g (per lb), therefore, the cost of the pure germinating seed is:

$$\frac{2 \times 100}{72} = £2.78$$

5
The Process of Germination

If we examine a seed in storage, for example a pea, we observe that there are no signs of life, yet the baby plant or embryo that the seed contains is very much alive and only requires the fulfilment of certain simple conditions to start it into growth.

Basically, the embryo consists of a plumule, which produces the shoot or top growth, a radicle which develops into the root and one or more seed leaves called cotyledons. With some seeds, on germination the cotyledons are carried above the ground. This is described as epigeal germination. Examples are brassicas, sunflowers and French beans. When the cotyledons remain below ground as with broad beans and peas this type of germination is called hypogeal.

Conditions for germination
Germination is bought about by allowing seed to have moisture, air and a suitable temperature. Perhaps the first is the most obvious requirement, for dryness is strongly associated with dormancy, and absorption of water by the seed is a necessary preliminary to germination. Some seeds, such as peas, beans, beetroot and carrots, may be soaked in water for a period before sowing. This enables them to absorb water more rapidly than they could in the soil and germination is accelerated. The majority of seeds, however, require only a limited steady supply of moisture, such as is present in moist soil, and are likely to be injured by being steeped in water.

The importance of oxygen for germinating seed is not always fully appreciated. Even when dormant, seeds respire slowly, but immediately they start to germinate respiration is accelerated to such an extent that it becomes greater at this phase of plant growth than at any other. From a practical point of view the gardener must avoid burying seeds so deeply in the soil that air is not freely available to them. Entry of air may also be restricted in wet, sticky soil, while heavy overhead watering may cause the soil to run together or cake on the surface, practically excluding air. Under such conditions the seed may be killed or give rise to poor weak seedlings.

If you sow seed outside in early spring, it germinates, but slowly. This is owing to the coldness of the soil and air. Some seeds, such as vegetable marrows, will very likely not germinate at all, for these require a minimum temperature of about 13°C (55°F) before they can germinate. On the other hand, mustard seed will sprout slowly when the temperature is near freezing point. This indicates the wide variation that exists in the temperature requirements of seeds for germination.

Up to a certain limit increasing the temperature accelerates germination; its effect is very obvious when seed is sown in a warm greenhouse. Each kind of seed, however, germinates best at a certain temperature, called the optimum. In general, a temperature which is about 5°C (10°F) higher than the optimum for the normal growth of a greenhouse-grown plant is most likely suitable for germinating seed of the same kind.

Seed should not be sown outside until the soil is warm enough to allow germination within a reasonable time. If, for instance, beet seed is sown in February it may decay before germination can occur. Other seeds (broad beans, parsnips and lettuces) may germinate successfully at that time. In greenhouses the temperature can be regulated to suit the particular kind of seed.

The germination of most seeds is not affected by light but there are some exceptions which are discussed later in this chapter. Light is nevertheless a vital necessity to all seedling immediately their leaves are spread out above the soil. If it is absent or not of sufficient intensity, seedlings remain pale and rapidly become weak and drawn up. For strong sturdy growth good light is essential.

Even when conditions are similar, seeds vary widely in the time they require for germination. Thus, willows have been known to germinate in twelve hours, mustard and cabbage in a day or so, while such vegetables as asparagus, parsnips and parsley may take several weeks. All these, however, respond to normal germination conditions and within a reasonable period after sowing they appear above the soil.

Delayed germination

The majority of seeds from cultivated plants germinate readily when subjected to the conditions described above. Unfortunately, there are many others which are much more difficult and may fail to grow for months or even years when given ordinary treatment. The principal causes of dormancy are as follows:

Immature embryos Some seeds when shed from the plant in apparently mature condition may require a period of what has been termed 'after-ripening'. Thus, freshly harvested lettuce seed is often slow to germinate but after a period in ordinary storage conditions a marked improvement occurs. Other examples of this type of dormancy include many species of orchids and ranunculus. In such circumstances apart from storage certain other treatments may be successful in breaking dormancy. If freshly harvested lettuce seed is kept in a refrigerator at 2°C (35°F) for 48 hours it usually germinates quite freely.

Hard seed-coats In certain seeds germination is hindered by hard seed-coats which prevent the entry of water and air. Many trees and shrubs have seeds of this kind and it is a common characteristic in both woody and herbaceous plants of the pea and bean family–the legumes.

Some of these (vetches and everlasting sweet peas) produce two types of seed known as 'soft' seed and 'hard' seed. The former has a seed-coat which is readily permeable to water, and germination is normal, but the hard seeds are impermeable and this restricts their germination.

There are various treatments recommended for seeds with hard coats. The simplest one is vigorous shaking, called impaction, which is claimed to be surprisingly effective for some legumes. The seeds may also be shaken in a jar or tin lined with sandpaper. Soaking the seed in alcohol with the object of dissolving the waxy layer of the seed coat has been used successfully with species of cassia. Another method is to chip the hard seed-coat with a knife until the seed itself is exposed and water can be taken up (Fig. 13).

More severe methods for hard-coat treatment have been used. One is to soak the seed in boiling water which, it is claimed, has successfully led to the germination of old abutilon seed. Another method is to immerse the seed in strong sulphuric acid, the time of treatment varying with the species. Thus hard seeds of clover require about thirty minutes, but brooms need a period of one hour in the acid and will tolerate up to five hours. Abrasion with sandpaper and chipping with a knife are common practices with certain hard seed such as lupins, cannas and sweet peas.

A more natural method for dealing with hard seed-coats is simply to bury the seed in moist soil or peat and keep it in a warm atmosphere. Under

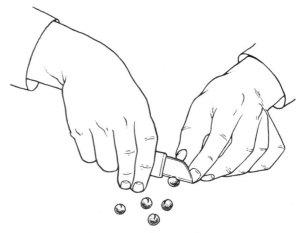

Fig. 13. Facilitating germination. Seeds with a hard seed coat can be chipped to expose the seed itself, thus allowing water to be taken up.

these conditions the hard seed-coats are decomposed by bacteria and fungi. Seeds that respond to such treatment include various species of cotoneasters and palms, *Halesia carolina* and nut-like seeds such as hazel.

Frost is also an effective means of rupturing hard seeds. The stony seeds of almonds, peaches, apricots and plums may be split by exposure to frosts, thus allowing the embryos to grow afterwards.

Stratification Many species of trees, shrubs and other plants may fail to germinate until they undergo a period of preparation in a low temperature and in moist conditions. This can be done by mixing the seed with peat and sand or by arranging layers of the seed and the materials alternately in pots or boxes. This treatment is called stratification (Fig. 14). In America much experimental work has been done on this

Fig. 14. Some seeds need to undergo a period of extreme cold before they will germinate. They should be placed in alternate layers with sand in a plant pot and stood outdoors during the winter. The process is known as stratification.

subject and tables have been prepared to show the time and temperature required for each species. A temperature of 5°C (41°F) is suitable for many kinds, and the period of treatment varies between two and four months. Apple seed, for instance, is treated for sixty to seventy days, at a temperature of 4.5°C (40°F).

Stratification can, however, be effected quite efficiently by simply leaving the containers, filled with seed and moist material, outside over

winter. Care must be taken to see that the seed is not eaten by rodents.

Seeds which benefit from stratification include a wide range of trees and shrubs. Many of these are found in the family Rosaceae, examples being cotoneaster, crataegus and stone fruits. Other examples are many conifers, acers, azaleas, daphnes, and magnolias.

Apart from rupturing hard seed-coats, frost has long been known to have considerable effect in breaking dormancy in many seeds, particularly alpines. With these a common practice is to sow the seeds in pans about Christmas and leave them exposed outside until February or March. Good results have also been achieved by freezing seed pans artificially in a domestic refrigerator or freezer. For this purpose a period of twenty-four hours freezing is said to be effective.

After being frozen it is usual to bring the seed into a greenhouse, or frame, or place it in a warm window. A temperature of 7°C (45°F) is high enough. If left outside, germination usually occurs later in the season. Of the many alpines which respond to this treatment, primulas, meconopsis and gentians are examples.

Light requirements

As previously mentioned the majority of seeds are indifferent to the presence or absence of light, but those which do have particular preferences may be placed in three groups:

(*a*) Those which germinate only in the dark. Examples are *Ailanthus altissima*, *Delphinium elatum*, *Euonymus japonicus*, *Hedera helix*, *Nigella damascena* and *Yucca aloifolia*.

(*b*) Seeds which require continuous or long periods of light, for example *Adonis vernalis*, *Bellis perennis*, *Genista tinctoria*, *Iris pseudacorus* and *Magnolia grandiflora*.

(*c*) Seeds which need brief periods of illumination only. Examples are *Nicotiana tabacum*, and, under certain conditions, lettuce.

Various theories have been advanced to explain the variability of seeds from different species in their response to light. Apparently very small seeds usually require light for germination and this is explained as a natural method of preventing such seeds germinating when deeply covered. Otherwise the tiny shoots would have little chance of reaching the surface. An interesting example are the extremely fine seeds of begonias which require at least three long day periods of at least 8 hours each before they can germinate.

The various factors which affect germination, such as temperature and light, cause further complications by interacting with one another. Thus an American variety of lettuce called 'Grand Rapids' immediately after harvest fails to germinate in the dark at 26°C (78°F) but when stored for a period will germinate in the dark at 18°C (64°F). At 26°C (79°F) it requires light stimulation.   Certain chemicals such as potassium nitrate and thio-urea are known to stimulate germination in some seeds. Thus seeds of several species when soaked in a 0–2% potassium nitrate solution showed a greatly improved rate of germination as compared with seeds soaked in water. Examples of such seeds are *Veronica longifolia*, *Hypericum hirsutum* and *Epilobium montanum*.

Natural methods

It is not surprising that the exposure of seed to low temperatures and frost when damp often gives good results, for nature uses this treatment widely. The seeds of many plants ripen in the autumn, fall to the ground and lie buried in moist soil or leafmould throughout the cold months of winter, and come to life in the spring. If they were to germinate in the autumn or winter the seedlings would probably be killed by frost.

On the other hand if the seed from such plants were collected, stored dry over winter and sown in the spring the results might be poor. A good example is *Acer palmatum* which is best sown as soon as ripe. The same applies to *Juniperus virginiana* and *Magnolia kobus*. All these require cold, moist pre-treatment.

Some seeds which have hard seed-coats also require stratification. This means, in effect, exposure for two periods in sequence, first to a high temperature and then to a low one, the seeds being kept moist all the time. A suitable high temperature for many seeds is about 25°C (77°F). This is followed by stratification at about 5°C (41°F) as previously mentioned. Seeds requiring this warm and cold treatment include cotoneaster, cornus, arctostaphylos, some species of berberis and soft-fruited conifers (junipers and yews).

A simple method of securing the benefits of high and low temperatures is to sow the seeds in question in summer when the warmth will promote decay of the seed-coats. The cold of the succeeding winter will stratify the seed and germination will probably occur in spring or early summer.

A further complication in germination is revealed with seed of *Lilium auratum*. These, if sown in a moderate temperature, produce roots readily but if a similar temperature is maintained development of the stems and foliage does not take place; in fact, experience has demonstrated that no further advance is possible until the half-developed seedling has experienced a period of cold treatment.

A practical method of dealing with this peculiarity is to apply the same remedy as advised for those seeds that possess hard seed-coats and also require stratification; that is, to sow the seeds in spring or early summer and allow them to pass through the correct sequence of temperatures necessary for complete development.

Other lilies such as *L. japonicum*, *L. rubellum* and *L. canadense* respond to the same kind of treatment as that advised for *L. auratum*. Similar conditions are also necessary for the seed of the ordinary herbaceous peony and the tree peony *P. suffruticosa* as well as *Viburnum dentatum* and *V. opulus*.

Conclusions

The examples given in this chapter indicate that there are many complexities in seed germination. A seed carefully harvested, apparently mature and seemingly perfect in structure, may fail to germinate under normal conditions until certain obscure changes in its physiology have been completed.

Yet nature often provides a guide to the most suitable method of treatment, particularly in relation to temperature. For instance, tomatoes, cucumbers, cosmos and zinnias, which have their origin in warm countries, usually germinate rapidly in high temperatures. Other plants that are native to colder climates may have an optimum germinating temperature that is appreciably lower. Moreover, most of these benefit from a kind of cold incubation pre-treatment or even freezing.

Many seeds germinate best when subjected to a particular sequence of high and low temperatures. A clue to such treatment is sometimes provided by a knowledge of the plant's habitat, time of seed production and natural germination. A considerable amount of research has been done on various aspects of this subject and is still proceeding. New and improved methods are constantly being revealed.

Keys to success

These considerations may act as a general guide to the gardener, but should he or she be in doubt about a particular kind of seed the following procedure can be advised. Seeds received in the spring should be sown at once in boxes or pans, placed in a greenhouse, frame or room. Mild heat is usually advantageous. Spring is the natural time for sowing many kinds, and if germination occurs within a reasonable period conditions are favourable for the growth of seedlings.

If early germination does not take place the seed should be exposed outside for a period during the winter and afterwards brought inside again.

When freshly harvested seed is available in the autumn a portion of it should be sown at once and the remainder in the spring. The autumn-sown lot may be kept indoors until about the New Year, and if there is no sign of germination should then be subjected to cold conditions. Treatment afterwards will continue on the same lines as for spring-sown seed.

During the waiting period frequent observation is necessary for signs of germination, and unless one has some knowledge of the usual time the particular kind requires for germination, no seed should be discarded for at least two years.

6
Seed Sowing in the Open

A considerable proportion of our garden crops and plants is raised by sowing seed in the open ground. These include most of the vegetables, many trees and shrubs, and a number of flowering annuals, biennials and perennials. A few plants successfully reproduce themselves from self-sown seed, examples being *Papaver alpinum*, erigerons, linarias, borage and limnanthes.

Sometimes the seed is sown where the plants are to remain and grow to maturity, in other circumstances the object is to produce seedlings for transplanting to another site. In the latter case, particular attention can be given to the selection of the most advantageous position for the seed-bed.

Preparing the seed-bed

Soil preparation should begin well in advance of sowing, for the ameliorating effect of frost on land turned over by spade or plough is better than any amount of cultivation. Moreover, land that is broken up in the autumn or early winter and left rough (Fig. 15a) allows the winter rains to enter and be reserved for the later use of seeds and plants instead of running off into the ditches.

A second phase in cultivation comes shortly before seeding. This consists in thoroughly pulverizing the soil with a fork or, more effectively, with one of the small mechanical rotary cultivators which are now so popular. Finally, the surface is levelled and broken down to a reasonable but not over-fine tilth with a rake, firmed by treading and raked once more to produce a fine tilth (Fig. 15b–d). During cultural operations the opportunity should be taken to remove perennial weeds. It goes without saying that land should never be cultivated when it is wet.

A very rich soil is not required for seedlings as it induces soft lush growth which is not desirable. For this reason heavy applications of farmyard manure or garden compost are not recommended unless the soil is very poor. In such circumstances well-rotted manure is preferable to fresh material which will rob the soil of nitrogen as it decomposes. Substances such as peat which lighten the soil and improve its water-holding capacity without unduly enriching it are beneficial.

Nitrogen (N), potash (K_2O) and phosphate (P_2O_5), are the three important plant foods and are required in balanced proportions. Nitrogen, however, promotes rapid growth, and fertilizers containing it should, therefore, be used judiciously for seedlings. The slower-acting organic forms of nitrogen, such as hoof and horn meal, are preferable for the purpose to the highly soluble fertilizers like sulphate of ammonia and nitrate of soda. Potash has no particular significance for seedlings, but as it is an essential plant food adequate supplies must be ensured.

Phosphate is particularly important for the early development of seedlings because it promotes root growth. Superphosphate is probably the best fertilizer in this class and is thoroughly recommended.

A suitable fertilizer mixture for seed-beds is as follows:

  4 parts by weight superphosphate
  1 part by weight sulphate of potash
  4 parts by weight hoof and horn meal.

This is applied at the rate of 55 to 85 g per sq m (2

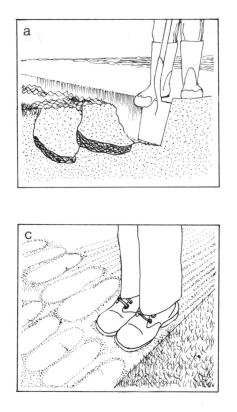

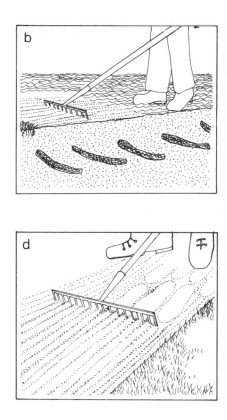

Fig. 15. Cultivation of outdoor seed bed for broadcast sowing. (*a*) Dig the soil to at least one spade's depth in autumn, incorporating compost or manure, if neces-sary. Leave the surface of the soil rough. (*b*) In spring rake the bed level. (*c*) Firm by treading. (*d*) Rake once more to produce a fine tilth.

to 3 oz per sq yd). On rich soil the hoof and horn could be reduced or omitted. Alternatively, a compound fertilizer with an analysis of 6%N, 8% P_2O_5 and 6% K_2O could be used at the same rate.

Lime is also extremely important in the raising of seedlings. The amount to apply depends upon the condition of the soil which should be tested with a simple kit to determine its lime content (known as the acidity or alkalinity of a soil).

Soil acidity is measured on what is called the pH scale. This scale runs from 0 to 14, but most soils in Britain have a pH value of between 4.5 and 8.5. Soils with a pH value of 7.0 are described as neutral. Those with a pH value above this are alkaline and those with a lower value are acid. Most plants prefer to grow in a soil with a pH value of between 6.0 and 6.5. Rhododendrons

and other acid lovers prefer a soil which has a pH of 5.5 to 6.0.

Soil testing kits will indicate the acidity or alkalinity of a given plot and will also recommend what action can be taken to adjust the reading to give a required pH level. Lime is alkaline and can be added in measured quantities to reduce acid-ity. Alkaline soil is more difficult to correct, though the addition of acidic fertilizers and organic matter such as peat does help.

How and when to sow
The correct time to sow seeds depends to a great extent on the kind of seed and when the plants are required to crop or flower. A primary considera-tion is to ensure favourable conditions both for germination and subsequently for the establish-

ment and development of the seedlings. Spring is naturally the most popular time, but several kinds may be sown during the summer and autumn, particularly in the months of August and September.

There are two methods of open-ground seeding; namely, sowing in drills and broadcasting. The former method is preferable for most seeds and is far more often used. It ensures better covering of the seed and allows for more convenient weeding and surface cultivation among the seedlings. Drills are excavated with an ordinary draw hoe, or triangular hoe drawn against a strong garden line (Fig. 16a-c). Begin on one side of the area to be drilled and stretch the line tightly from one end of the plot to the other. Take particular care to get the first drill running in the correct direction and not at a peculiar slant. The drill is drawn with the corner of the hoe, allowing the blade to run close to the line, the operator walking backwards, preferably on the line to prevent its being moved by the hoe.

For even germination the drills must be level at the bottom, and their depth will depend on the type of seed to be sown. Similarly, the distance between the rows varies according to the crop and may be from 15 to 45 cm (6 to 18 in), but 30 to 38 cm (12 to 15 in) is most usual. Careful measurement at each end of the row is essential as the line is moved over for each new drill, to ensure even spacing.

Sowing technique
Sowing is done by holding a small quantity of seed in the hand and allowing it to fall in a steady even trickle by the movement of the fingers and thumb (Fig. 16d). Thin sowing is usually advised, but sufficient must be allowed to ensure an even stand of plants without gaps. A method of making seed go further is to sow three or four seeds together in groups which are spaced the correct distance apart for mature plants. The seedlings can later be thinned to leave one plant at each station. After sowing, fine soil is drawn over the seed with the back of the rake, and the surface made level (Fig. 16e).

A wide range of seeds may be sown by a mechanical drill, e.g. cabbage and peas, but it is necessary to adjust the outlet in accordance with the seed size. A line is necessary for the first row to ensure straightness, but afterwards may be dispensed with as most drills are provided with a device to mark the position of the next row and the machine may be adjusted for different spacings.

Broadcasting is used only for a few kinds of seed, such as hardy annuals sown where they are to flower, and grass seed for a lawn. Shortly before sowing, the ground is raked absolutely level and very fine. Finally, draw the rake firmly through the soil so as to leave the surface as a continuous series of small drills. Over these the seed is scattered thinly and evenly by hand and is then covered by raking in several directions. Grass seed for lawns should, however, be sown very thickly to ensure a good sward. To provide better cover it is often an advantage to sieve or scatter by hand a little dry soil over the sown area. This is particularly valuable if the seed-bed soil is somewhat wet and sticky; in such circumstances it may be advisable to cover seed sown in drills similarly.

The correct depth to cover seed depends upon their size. Fine dustlike seed such as poppy should hardly be covered at all; turnips, carrots, leeks and onions require about 1 cm ($\frac{1}{2}$ in) cover; radishes, parsnips and beets 2.5 cm (1 in); peas and lupins 5 cm (2 in), and runner beans 8 cm.

In early spring rather shallow sowing may accelerate germination as the surface soil warms up first, but in summer deeper sowing may be advisable to ensure a more reliable moisture supply. Sowing too deeply, however, may starve the germinating seed of oxygen as well as hindering and delaying the seedlings in reaching the surface. As a general rule a seed should be covered to one-and-a-half times its own diameter or length.

After sowing it is usual to firm the soil over the seed unless the ground is wet. This is done by treading, or by patting the soil with the back of a rake. Consolidating the surface in this way brings the seed in closer contact with the soil particles and the moisture that surrounds them.

Watering the seed-bed after sowing is rarely advisable as it may wash the soil off the seed and also causes caking of the surface. If the land is very dry a good soaking should be given before sowing and watering is often advantageous when the seedlings are well through. Always give enough water to moisten the top few inches of soil and apply with the finest possible spray.

Precision drills
Precision drills are widely used on farms and market gardens. These are power-operated but

Fig. 16. Sowing seed out of doors in drills. (*a*) Dig the soil deeply in autumn. In spring break down the surface of the soil, firm by treading and rake to a fine tilth. (*b*) Stretch a line taut across the soil to mark the position of the drill. Push the stake at the end of the line into the soil at an angle so that as the stake is pushed down the line tightens. (*c*) Take out a drill using a draw hoe with the blade resting on the line. Use your foot to prevent the line moving.(*d*) Sow seed thinly. (*e*) Mark the ends of the drill and cover the seeds by raking soil over them from both sides of the drill.

hand-pushed models are also available for the amateur. These drills will sow single seeds or small groups of 2 or 3 seeds at a predetermined spacing and depth. A wide range of seeds from cabbage to peas and beans as well as flower seeds and pelleted seeds of all kinds can be sown in this manner. Some models however, cannot deal with seed above the size of beetroot seed.

The advantages of seed spacing is a saving of seed and a reduction in labour for many crops. Thus, spaced seed may require little or no thinning and transplanting may be obviated. Cabbage, cauliflower and lettuce can, for instance, be space-sown *in situ* without transplanting.

One disadvantage of precision drilling is that the reduced number of seeds used necessitates a high rate of germination without which a poor plant stand is likely. Consequently, extra care is needed to ensure a good, moist, well-prepared seed-bed for drilling. It is also most important to maintain surface moisture by frequent fine but thorough sprayings during dry weather.

Pelleted seed

This is a method of coating seeds with a layer of inert substance, thereby increasing their size, rounding them and thus facilitating handling. Pelleting is in fact complementary to precision sowing because it allows these drills to handle a wider seed range including awkwardly shaped

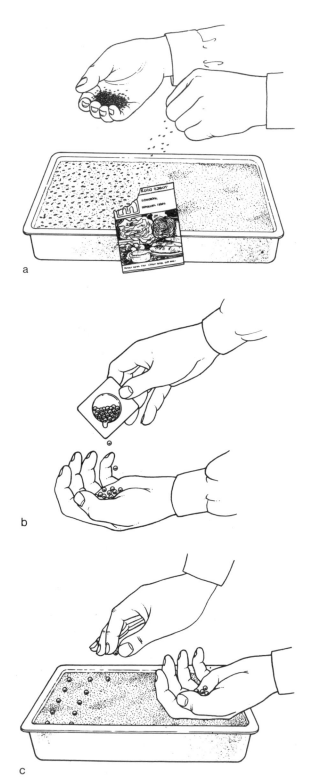

a

b

c

seed such as onions and lettuces. Some pellets contain a fungicide to protect the seed during germination.

The spacing of pelleted seed by hand is also quite feasible for the amateur. Thus, many treated vegetable seeds and hardy flower crops can be conveniently spaced in the open with a consequent saving of seed and labour. Similarly, pelleted seed can be easily spaced in seed boxes and pots so that pricking-off is obviated (Fig. 17). A wide range of pelleted seeds is available from seedsmen including very small flower seeds like alyssum and antirrhinums.

Pelleted seed requires special care in storage as the pellets tend to absorb moisture. They should therefore be kept in a tightly sealed plastic container. Extra good conditions are also necessary for satisfactory germination and it is important to maintain the surface soil in a moist condition. Even temporary dryness may cause complete failure.

Fluid sowing

This new seed sowing technique has been developed in recent years and it is particularly useful with seeds that are slow to germinate, or where the soil does not offer ideal conditions for germination.

With fluid or liquid sowing, the seed is pregerminated or 'chitted'. Moist tissues or blotting paper are placed in a shallow dish or plastic sandwich box, the seeds are sown thinly and evenly over this moist pad (Fig. 18a) and the container is then covered and placed in a warm cupboard or other suitable place where a temperature of around 16 to 18°C (60 to 65°F) can be maintained.

As soon as the seeds have started to produce small white roots they must be sown. Do not let these roots grow to more than 1 cm ($\frac{1}{2}$ in) or they may be damaged in the sowing operation.

Any ordinary wallpaper paste that does not contain fungicide or other impurities is a suitable medium for fluid sowing. Mix the paste up to half

Fig. 17. Pelleted seed versus non-pelleted seed. Note that pelleted seed can be spaced with precision in the seedbox (*b* and *c*), unlike the non-pelleted kind (*a*). However, with pelleted seed greater care is needed to ensure maintenance of correct moisture level, for failure is common if the soil dries out during germination.

its recommended strength (using either half the amount of powder or twice the amount of water) and then carefully mix, to ensure an even distribution, the germinated seedlings into the paste.

A seed drill is taken out on the plot in the normal fashion, but do make sure that the soil is moist. Pour the paste/seed mixture into a polythene bag, twist-seal the top and then cut off a small corner of the bag to make a nozzle through which the seeds can be squirted into the drill rather like icing or toothpaste (Fig. 18b). The soil can then be drawn over the seeds in the usual way.

Germination of chitted seeds sown in fluid is very rapid and the young plants get away to a flying start, but care must be taken to ensure that the soil does not dry out in the weeks immediately after sowing.

Chitted seeds

Some seed merchants are now selling pregerminated or 'chitted' seeds of certain crops that are difficult or costly for the amateur gardener to raise at home. Cucumber seeds need a lot of heat in the early stages and the seed of F_1 hybrids can be very costly. Packs of five or six seedlings ready for pricking out have recently been made available by one firm, and such young plants do stand the greatest possible chance of succeeding if potted immediately they arrive. They may appear more costly per plant than a packet of seeds, but their survival is almost guaranteed and may represent a saving in the long run.

Weed control

There are three methods of keeping weeds in check: (a) by soil cultivation and hand weeding; (b) by chemical herbicides; (c) by mulching.

When the ground is first cultivated every effort should be made to get rid of perennial weeds such as couch grass and bindweed. Although many of these do respond to herbicides there is much to be said for a careful initial forking out and removal of the roots of such plants.

As regards basic cultivations, weed control, particularly of annual weeds, is best effected by turning the soil over a spit deep by spade or plough. This is because burying the surface soil to a reasonable depth retards or prevents a high proportion of weed-seed germination. Rotary cultivation on the other hand leaves weed seeds

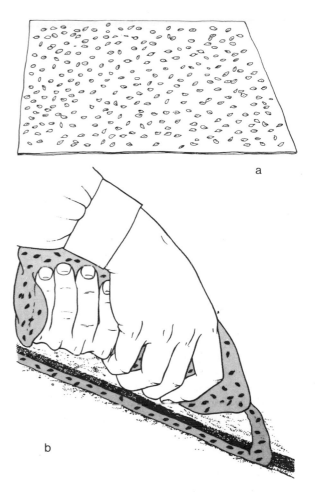

Fig. 18. Fluid sowing. (a) Seed is germinated on moist paper tissue, which is placed in a warm environment such as an airing cupboard. (b) The germinated seed is mixed with paste as soon as small roots can be seen and poured into a polythene bag. The mixture is squeezed through a hole in the bag into the seed drill.

near the surface thereby promoting their germination. Weed plants after rotavation are also left near the surface and, unless removed quickly, re-establish themselves.

The Dutch hoe is perhaps the most useful weeding tool in the garden, for it can be pushed through the soil between plants in rows or beds so that its blade travels just below the surface. In this way it will sever the stems of annual weeds from their roots, causing them to die. It will not have any effect on perennial weeds which must be removed completely or else treated with a herbicide.

Despite the wider use of herbicides a good deal of hoeing and hand-weeding is still necessary in most gardens. It is important that this work is done sooner rather than later for weed competition has a serious effect on the growth of seedlings. Also, when weeds are removed early, seeding and consequent reinfestation may be prevented.

Chemical weed control

In recent years great advances have been made particularly with vegetable crops in the control of weeds by herbicides. This method, however, should still be regarded as complementary to cultural operations, and for a few specific crops only, may be adequate on its own. Furthermore, owing to spray drift and other factors it is difficult to treat a particular crop in a congested garden without inadvertently damaging other susceptible crops or plants. If herbicides are used in a garden the treatment should always be done on a calm day and the diluted herbicide applied with a watering can fitted with a dribble-bar (Fig. 19). This will avoid 'spray drift' which would damage nearby plants.

Fig. 19. Weed control. The use of a dribble-bar attachment on the watering can prevents spray-drift on to nearby crops.

From the practical point of view herbicides may be placed in three groups:
1. Hormone or translocated herbicides.
2. Contact herbicides.
3. Soil-applied residual herbicides.

Translocated herbicides are certainly the most dangerous to use, for if just one leaf of a cultivated plant is wetted by the solution the entire plant may suffer. However, on ground which is being cleared of weed growth, and where no cultivated plants are being grown, such herbicides are very useful. They may be lightly watered over the weeds with a watering can fitted with a fine rose or a dribble-bar. Take care to keep one can specifically for herbicides. The residue will be difficult to wash out completely and may harm cultivated plants that are watered some time later. Glyphosate is a translocated herbicide that offers control over a wide range of perennial weeds including couch grass and ground elder.

Contact herbicides which kill or damage plants on contact can be divided into two groups:

(*a*) Those which are toxic to most plants and crops and therefore can be used only when crops or garden plants are not present or, if they are present, the herbicide must be directed on to the weeds only. Good examples are paraquat and glyphosate.

(*b*) Selective contact herbicides. These destroy most weeds but for certain specified crops have a selective action. On these they may be used as an overall spray taking care that there is no drift on to adjacent susceptible crops. Ioxynil is a good example and can be used for weed control on lawns without harming the grass.

Soil-applied residual herbicides are used on soil which is largely weed-free at the time of application. Such herbicides usually persist in the soil surface for a period and kill off weed seedlings as they germinate. Herbicides of this type are often applied to a seed-crop after sowing but before the crop seedlings emerge. They may also be applied to growing crops. Residual herbicides are largely selective in action and cannot be used indiscriminately on all crops. Several other residual herbicides are useful for weed control among roses and other shrubs which are relatively deep rooting. Simazine and dichlobenil are examples. A chemical used as a pre-emergence herbicide among vegetables is propachlor (available as Herbon Orange). Further advice and useful leaflets on the subject of herbicides can be had from the Ministry of Agriculture, Fisheries and Food (Publications), Tolcarne Drive, Pinner, Middlesex HA5 2DT.

All gardeners would be well advised to arm themselves with a copy of the Directory of Garden Chemicals, published by the British Agrochemicals Association, Alembic House, 93 Albert Embankment, London SE1 7TU. This booklet shows exactly which chemicals are available to the gardener, and the trade names under which they are sold.

Mulching

The placing of a layer of organic material such as peat, pulverised bark, leafmould, well-rotted manure or garden compost on the soil around ornamental shrubs, fruit trees and bushes, is known as mulching.

The mulch should be applied to clean, moist ground. As well as suppressing weed growth it will also help to conserve moisture and enrich the soil.

Around strawberries a mulch of black polythene will fulfill the first two requirements.

Organic mulches are best applied in spring and dug into the soil in autumn. For best results they should be 5 to 8 cm (2 to 3 in) thick.

Thinning and transplanting

When seedlings are not space-sown they usually come up too closely together and must be either thinned out or transplanted suitably spaced to allow the maximum development of the plants. Some species, such as vegetable root crops, are difficult to transplant and are sown where they are to remain and mature. Thinning may be done in two operations. On the first occasion the weakest plants are removed and the others left about twice as thick as required. At the second thinning where all have survived, alternate plants are pulled (Fig. 20a). This method allows for casualties during the early stages of growth.

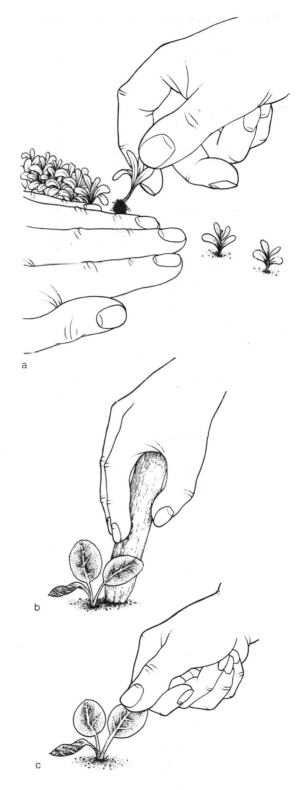

Fig. 20. Thinning and transplanting. (*a*) To thin seedlings pull out the unwanted ones with one hand while pressing down the soil with the other hand to prevent other seedlings being disturbed. (*b*) Firm a transplanted seedling by levering soil against its roots with a dibber. (*c*) Test the firmness of planting by pulling a leaf between finger and thumb – it should tear rather than pull up the plant.

49

Transplanting should be done as soon as convenient, for it is found that small seedlings suffer less check from being lifted than larger ones. It is often an advantage to soak the seed-bed the day before lifting and to ease up the seedlings with a fork before pulling.

The seedlings should be transplanted in well-prepared soil and preferably in dull, showery weather. The roots should be kept covered during the period the plants are out of the soil. The tools for transplanting are the dibber and trowel and a garden line to ensure straight rows. A dibber is quite suitable for the speedy planting of small seedlings, for example those of the cabbage tribe and lettuce. Older plants, however, which have developed a mass of fibrous roots are best planted with a trowel.

To plant with a dibber it is necessary to hold it in the right hand and the plant in the left. Make a hole with the dibber and place the plant in this, taking care that the roots are not turned upwards. Insert the dibber again an inch or so from the plant and then use it to press the soil firmly against the stem (Fig. 20b and c). In dry weather, for some plants (brassicas for example), it is a good plan to take out a drill and set the plants in this. The drill can be flooded after planting.

When using the trowel drive it into the ground vertically and pull the soil out towards yourself. Insert the plant against the back of the hole and replace the soil firmly against it. Special planting machines are available to plant large numbers of plants, and very often the results are better than when the plants are set by hand.

After transplanting, a good watering is often beneficial if the soil is dry. Transplanted seedlings also benefit from a light spray in the evenings of hot days. In bright sunny weather shade may be provided by lath screens supported above the plants on pegs 30 to 60 cm (1 to 2 ft) high. Another device is made by stretching wire netting tightly on a wooden frame which is supported over the plants. Brushwood is then laid on the wire to provide shade. The screens should not be kept in position longer than necessary, otherwise the plants may become drawn up. Screens of this type also afford protection against birds, but for this purpose fine nylon or plastic netting is quite effective.

7
Seed Sowing under Glass

By sowing seed in greenhouses and frames the perils of weather and from various pests are to a considerable extent eliminated, for the gardener can exercise far more control over the germinating seeds' environment than is possible out of doors. He can, in a very real sense, create his own climate to suit the type of plants he is propagating and, particularly in a heated greenhouse, a higher and more uniform temperature can be maintained.

Naturally, these conditions result in speedier germination and the more rapid development of the seedlings. Thus, even with plants that are quite hardy in this climate there are advantages in sowing under glass. This is very true when raising plants from relatively small quantities of seed, perhaps some of which are of special value. Moreover the seed of certain hardy plants, such as rhododendrons and ericas, is very fine and requires extra care when sowing.

Seed-sowing in greenhouses also enables one to raise a number of vegetables extra-early in the season. These are planted outside when the weather is warmer and result in earlier crops or, due to the longer period of growth, higher yield. Examples of vegetables treated in this way are lettuces, brassicas, onions and leeks.

Many half-hardy plants must be raised under glass and are planted outside when the danger of spring frosts is over and growing conditions are more favourable. Starting such plants under glass also allows more time for their growth and development. Vegetable marrows, tomatoes and many half-hardy annual flowers provide examples. True greenhouse plants like melons and gloxinias must, of course, be propagated under glass.

Many seeds may be germinated in a warm room, but they must be moved to a well-lighted window immediately they are through the soil.

Seeds and seedlings raised indoors escape the ravages of many garden pests, but to minimize loss or damage from greenhouse pests and disease strict hygiene is most essential. All greenhouses should be washed down occasionally with a diluted solution of horticultural disinfectant such as Jeyes' Fluid or Clean-Up. No rubbish should be left lying about to provide cover for pests and act as a starting place for disease. Routine fumigation is usually necessary to keep down such common pests as greenfly and thrips. Water can also be a means of infecting plants with disease, and great care should be taken to ensure a clean supply.

Seed and potting composts
Outdoor plants grow in ordinary soil, but in greenhouses seedlings are usually raised in a special mixture called a compost. At one time it was thought that almost every plant grown in a container required a different mixture, but experiments at the John Innes Horticultural Institute in the 1930s demonstrated the success of growing the majority of such plants in the same mixture. Hence, two standard composts were devised, John Innes Seed compost, usually abbreviated to JIS, and the John Innes Potting compost, known as JIP. The former is used for seed-sowing and the latter for potting and growing plants on to maturity.

These composts consist of three bulky ingredients and concentrated fertilizers. The bulky materials are loam, coarse sand and moss peat.

The loam is normally secured from the top 15 cm (6 in) or so of good pasture land and should be moderately heavy. The turves are usually cut and stacked grass-side downwards for a period of 6 to 12 months, and are prepared for use by chopping them up finely with a sharp spade. A soil shredder does the work quickly and effectively. The soil is then riddled through a 1 cm ($\frac{3}{8}$ in) sieve. It is a good idea to have a sample of the loam chemically analysed and if any plant foods, for example, phosphate or lime, are in short supply, the deficiency should be made good before mixing.

Coarse, clean sand or washed grit with the particles up to 3 mm ($\frac{1}{8}$ in) in diameter is necessary. This is to aerate the compost and provide drainage. Fine sand has very little effect.

Moss or sedge peat is preferred, and very fine dust-like material should be avoided. Moss peat is usually procured in bales and should be broken up by the soil shredder or by other means, and the peat should be well moistened before being mixed with the other ingredients.

The standard composts are made up as follows:

John Innes Seed Compost (JIS)
2 parts by bulk loam
1 part by bulk moss peat
1 part by bulk coarse sand or grit.

To each bushel of this mixture add 40 g (1½ oz) superphosphate (18% phosphoric acid), 21 g ($\frac{3}{4}$ oz) chalk.

John Innes Potting Compost (JIP)
7 parts by bulk loam
3 parts by bulk moss peat
2 parts by bulk coarse sand or grit.

To each bushel of this mixture add 110 g (4 oz) John Innes base fertilizer and 21 g ($\frac{3}{4}$ oz) chalk.

J.I. base is available from horticultural sundriesmen but can be prepared at home by mixing together:

0.9 kg (2 lb) hoof and horn (3 mm ($\frac{1}{8}$ in) grist) (13% nitrogen)

0.9 kg (2 lb) superphosphate (18% phosphoric acid)

0.45 kg (1 lb) sulphate of potash (48% pure potash)

The John Innes Potting Compost with a single dose of fertilizers as shown above is known as JIP₁. If a double quantity of fertilizers is added, that is 225 g (8 oz) J.I. base to the bushel, the compost is distinguished as JIP₂, and similarly if the fertilizers are trebled then the compost is JIP₃. Where the fertilizers are doubled or trebled the chalk should also be increased in the same proportion.

Seed and potting composts should be thoroughly mixed and sufficient may be prepared at a time for several weeks use. Long storage, however, is not advocated.

Sterilizing the soil
Normally, soil used for making the John Innes compost is given special heat treatment often called sterilization; in effect it achieves only partial sterilization. The object of soil sterilization is to destroy injurious plant pests like wireworms and eelworms. This process also releases plant foods and thereby improves fertility.

In large nurseries soil may be sterilized by placing it in special bins which are provided with a grid of perforated pipes at the bottom. The grid is connected to a boiler from which steam is forced through the pipes into the soil for a sufficient period to heat the soil to a temperature of 82°C (180°F) and maintain it at that for about 10 minutes.

There are various methods of sterilizing small quantities of soil. Perhaps the simplest is to put about 1.3 cm ($\frac{1}{2}$ in) of water in a large saucepan and then fill up loosely with dust-dry soil. Bring the saucepan to the boil and allow to simmer for about 15 minutes. Afterwards the soil is spread out to dry on a perfectly clean surface.

Electrical soil sterilization (Fig. 21) is probably the most convenient method for the private gardener. Sterilizers of this type treating up to a bushel of soil at a time should be suitable in most gardens. The instructions provided with these are easy to follow and current consumption is not heavy.

Whatever method of sterilizing is used, the temperature should be raised to the required point as quickly as possible, but kept at that for no longer than is necessary. Everything possible should be done to ensure uniform heating in the soil, avoiding 'cold spots' and local overheating.

Chemical sterilization offers an alternative to heat and although not quite so effective does

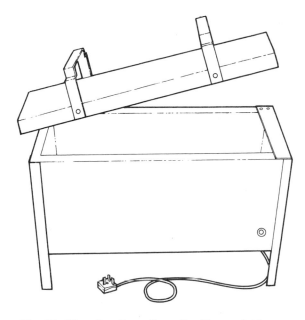

Fig. 21. Electric soil sterilizer. Sterilizers of this type can treat up to a bushel of soil at a time.

ensure a fair degree of freedom from soil-borne fungi such as the damping-off disease of seedlings. It should also destroy weed seed and soil pests.

To work effectively chemical sterilization involves the maintenance of a warm environment over the period of treatment. In this respect and in all others the maker's instructions should be followed. Usually the loam to be sterilized is broken up fine and a 23 cm (9 in) deep layer of it spread on a clean surface such as a bench or concrete floor. The chemical is then added and thoroughly mixed with the soil according to instructions. The most useful substance now available is Dazomet which is applied as a dust and is therefore convenient for the amateur.

After treatment the soil should be kept covered with plastic sheets for the recommended period of time. Subsequently, the covering should be removed and free ventilation given. Turning the sterilized soil will promote the clearance of toxic fumes. It is usual to allow a period of 4 to 6 weeks before the soil can be used in a compost for seed sowing.

Particular care must be taken with sterilized soil to prevent contamination. No contact with unsterilized soil should be permitted, while mixing floors, places of storage and soil implements should be kept scrupulously clean. Sterilized soil should preferably be kept in a bin or other container fitted with a lid.

All receptacles used for propagation should be scrupulously clean before being filled. If they have been used before they should be washed and sterilized. Boxes, or trays and pots made of plastic, may be sterilized by dipping in a formalin (formaldehyde) solution prepared by mixing one measure of 40% formaldehyde with 49 measures of water. This gives a one in fifty dilution. Afterwards the trays should be exposed to the air and thoroughly dried before use. Clay pots, being porous, are liable to retain some of the chemical with consequent damage to plants growing in them later. These should preferably be sterilized by placing them in water which is then brought to the boil. Crocks for drainage should also be washed and sterilized.

Gardeners today have an opportunity for dispensing with pots in certain circumstances by using soil blocks instead. These are made from ordinary compost with a special implement which presses the material together to form a block somewhat similar in shape to a pot. Compost for block-making should be fairly moist. The finished block has a small hole centrally placed at the top, in which the seed or seedling is planted with a little soil.

Soil blocks should be prepared shortly before being required. They should never be allowed to dry out and are kept moist by spraying with water both before and after the plants are inserted in them. They serve a similar purpose to small pots and are mostly used for the stronger growing seedlings which are later planted out in beds or borders or set in large pots.

Several manufacturers now produce peat-based blocking composts together with very strong plastic moulds (Fig. 22). They are used in the same way as soil-based composts and given similar aftercare. Pay particular attention to watering and do not allow the blocks to dry out.

A compromise between peat blocks and ordinary plant pots can be obtained if compressed peat pots are used. These are filled with compost and kept constantly moist. Seeds may be individually sown in peat pots, or seedlings pricked out into them. They are most easily managed if placed on a capillary bench (see p. 25).

Compressed peat pellets (marketed as Jiffy 7's)

a

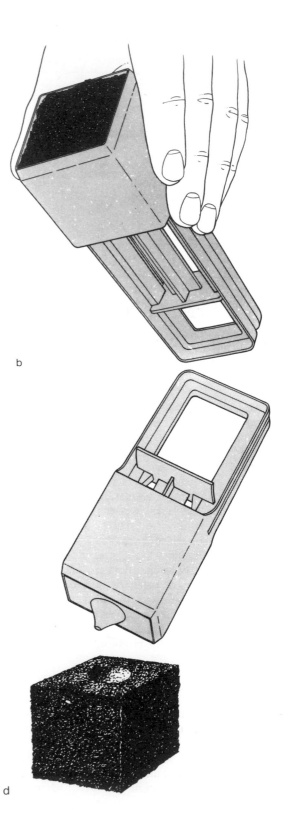

b

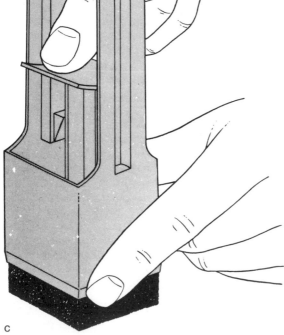

c

d

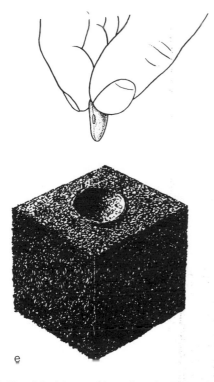

Fig. 22. Peat-blocking tool in action. (*a*) Take a handful of moistened blocking compost. (*b*) Fill the tool with compost. (*c*) Invert and press down lever. (*d*) Remove tool. (*e*) Sow seed and cover it with compost. Keep blocks moist at all times.

are also useful for direct sowing. The pellets are about 3 cm (1¼ in) in diameter and when soaked in water will expand into cylinders of peat compost contained within a plastic netting that allows roots to penetrate through it. These pellets are illustrated in Fig. 23.

The advantage of using peat pots and pellets is that plants being grown in them can be planted out or potted on with very little disturbance, for the containers do not need to be removed. Take care to ensure that they are never allowed to dry out, and take the precaution of removing the plastic net from the Jiffy 7's at planting time to make sure that roots are not restricted.

Soilless composts
During recent years there have been considerable developments in the use of loamless or soilless composts. These consist of mixtures of peat and sand or of pure peat used alone. Other mixtures comprise peat and vermiculite or peat and perlite.

The advantage of soilless composts is a higher degree of uniformity due to less variability in their components than in the case of loam-based composts. They are also lighter in weight, cleaner to handle more convenient to store and do not require sterilization. The main difference between loam-based and soilless composts is that the soilless types should never be allowed to dry out completely, they need less firming and they will become exhausted of fertilizers more rapidly than their soil-based equivalents. However, as far as seed sowing is concerned they often produce far better results than John Innes composts.

Today there are several proprietary soilless composts on the market which can be recommended to the amateur. Some of these are pure peat, others are peat and sand mixes. As a rule one should select a brand which is available as (*a*) a seed compost and (*b*) a potting compost. It is asking a lot to secure a compost suitable for both purposes. Some proprietary composts appear to be more suitable for certain plants than for other kinds. Hence the gardener should experiment until he finds one which suits his purpose best.

The amateur can also try making his own soilless compost which, if properly done, may be equal to the best of the proprietary brands. Many years ago the University of California developed a wide range of soilless composts which have been used successfully in many parts of the world. An example of a University of California potting compost is as follows:

 3 parts peat (2 parts in winter)
 1 part fine sand.
To each 0.19 cu. m (¼ cu. yd; approximately 5½ bushels) add:

 30 g (1 oz) potassium nitrate
 285 g (10 oz) superphosphate
 30 g (1 oz) sulphate of potash
 285 g (10 oz) chalk or ground limestone
 850 g (30 oz) magnesium limestone.
Perhaps the most suitable composts for British conditions are those devised by the Glasshouse Crops Research Institute at Littlehampton. The ingredients comprise sphagnum moss peat and lime-free sand having a particle size range of 0.05 mm to 0.5 mm which is much finer than the sand specified for J.I. composts.

The formulae are as follows:
 1. For seed sowing:
Equal parts by volume of peat and fine sand. Add to each bushel:

14 g ($\frac{1}{2}$ oz) sulphate of ammonia
30 g (1 oz) 18% superphosphate
14 g ($\frac{1}{2}$ oz) sulphate of potash
110 g (4 oz) ground chalk or limestone.

2. For potting and pricking out in boxes:
Three parts by volume of peat and one part of fine sand.

Add to each bushel:

I: for immediate use or short-term storage only:
7 g ($\frac{1}{4}$ oz) ammonium nitrate
14 g ($\frac{1}{2}$ oz) urea/formaldehyde (38% N)
30 g (1 oz) sulphate of potash
55 g (2 oz) 18% superphosphate
85 g (3 oz) chalk or ground limestone
85 g (3 oz) Dolomitic limestone
14 g ($\frac{1}{2}$ oz) Frit No. 253 A.

II: For longer term storage:
14 g ($\frac{1}{2}$ oz) ammonium nitrate
30 g (1 oz) potassium nitrate
55 g (2 oz) chalk or ground limestone
85 g (3 oz) Dolomitic limestone
14 g ($\frac{1}{2}$ oz) Frit. No. 253 A.

Frit No. 253 A supplies trace elements and is manufactured by Ferro Corporation, Wombourne, Wolverhampton, and can be purchased only through Tennant Trading (Chemicals) Ltd., 141 Rainsford Road, Park Royal, London NW10 7SD.

Seed sowing in pots and trays

The filling of receptacles with compost for seed-sowing should be done carefully. Clay pots larger than 8 cm (3 in) diameter should have clean, broken crocks placed concave side downwards in their bases to prevent the single drainage hole from becoming blocked. The compost may then be put in, firmed with the fingers and finished off level, with the soil surface about 2.5 cm (1 in) below the top of the pot. For small seed, finish off by sieving a little fine compost over the top. With begonia seed it is a good idea to dust a little fine silver sand over the surface. This enables the sower to see the darker seed and helps to ensure even distribution.

A wide variety of pots and trays for seed sowing and propagation is available to the amateur gardener today, some of which are shown in Fig. 23.

Seed trays for raising quick-growing seedlings are filled directly with compost, no drainage material being provided. The same goes for plastic pots and for any containers in which soilless compost is to be used. Trays are filled with compost to within about 2.5 cm (1 in) of the top. The soil should be pressed down firmly with the fingers, (lightly if soilless compost is used) particularly at the corners; afterwards the surface is made level by lightly pressing with a firmer or patter. For slow-germinating seed a material such as coarse gravel may be laid in the bottom of the tray.

If there is a quantity of seed to sow of various kinds it is advisable to prepare the requisite number of containers in advance. Each of these is then thoroughly soaked with water and left for a while to drain. Next, write labels for each container to show the variety or kind of seed to be sown in it. The labels may be pushed into the compost or laid at the back of each container.

Sowing should be done carefully, using the finger and thumb in preference to sprinkling directly from the seed packet. Aim to sow evenly and thinly. Thick sowing often gives rise to a mass of weak seedlings, and there is a greatly increased risk of the dreaded damping-off disease wiping out the seedlings in patches. Covering is done by sieving a little compost over the seed, the depth depending on the type. About 3 mm ($\frac{1}{8}$ in) is sufficient for seeds such as lettuce, cabbage or onion. Larger seeds like those of tomatoes should be covered to a depth of about 5 mm ($\frac{1}{4}$ in). Dust-like seed, such as that of begonia, lobelia, gloxinia, and streptocarpus, needs no covering at all.

It is not advisable to use very fine soil for covering large seed; thus tomato seed does best when covered with soil sifted through a 5 mm ($\frac{1}{4}$ in) sieve; with lettuce and onions a 3 mm ($\frac{1}{8}$ in) sieve should be used. Very fine soil tends to cake on the surface, thus excluding air.

After sowing is completed it is usual to place all the containers together and cover them with sheets of glass and paper. This is to prevent drying out and to maintain an even temperature. The receptacles should be examined daily for signs of germination, and at the same time condensed moisture wiped off the glass. Immediately germination occurs the seedlings must be placed in full light. Should they be left covered for a few days they may be worthless. Subsequently, good light conditions must be allowed. Seedlings grown in shade soon become weak and drawn up.

On the other hand, young plants under glass may be severely damaged by a period of bright

Fig. 23. Pots and trays for propagation. (*a*) Clay pots. (*b*) Standard-size seed tray with plastic top – note the two ventilators. (*c*) Standard and half-size trays. (*d*) Plastic pots. (*e*) and (*f*) Jiffy 7's. (*g*) Round and square compressed peat pots. (*h*) 'Whalehide' pots. (*i*) Peat strip pots. (*j*) Polystyrene slab pots.

sunshine and the gardener must beware of such a possibility. Usually seedlings can be protected from strong sunshine by covering them temporarily with sheets of newspaper, but these should not be left on longer than is absolutely necessary. Some greenhouses are provided with roller blinds which are ideal for this purpose. Permanent shading provided by whitewash is not advisable during the early part of the season, but is practicable in summer.

If the seeds are sown on moist compost and germination occurs within a reasonable period, watering may not be necessary until the seedlings are above the soil. If the compost dries out before germination the receptacles should preferably be carefully dipped in water and allowed to soak up moisture. In nurseries, however, this method may be too slow and it is necessary to resort to overhead watering. This should be done carefully with a fine rose to avoid washing the soil off the seed. After germination water should be given when required, and in hot weather may be necessary at least once a day.

Pricking out

The next stage in seed propagation is to transplant the seedlings or, as it is usually called, pricking out or pricking off. Seed trays may be

Fig. 24. Transplanting seedlings. Seedlings should be transplanted (pricked out) from their original pot or tray as soon as they can be handled. Use a dibber to make holes in the compost in the seed tray large enough to take the roots comfortably, insert the plant and lightly firm the soil.

used for this purpose. Fill them with compost in the same way as for seed sowing, but use John Innes No 1 potting compost or its soilless equivalent. Plants such as tomatoes may be pricked out individually into small pots.

Transplanting seedlings (Fig. 24) is necessary to allow them space for development, but it does cause a check to growth; hence, the fewer moves a seedling has the better. Experiments have shown that the younger a seedling is the less check it suffers on pricking out. It is, therefore, advisable to prick out seedlings immediately they can be handled. Tomatoes, for instance, should be transplanted the second day after they appear above the soil.

Seedlings should be lifted carefully and every effort made to reduce root damage to the minimum. The average spacing is 4 to 5 cm (1½ to 2 in), but 8 cm (3 in) may be necessary for the stronger-growing kinds. Spacing should be even and the rows kept straight.

After pricking out seedlings are provided with a humid atmosphere and may require shading for a day or two. This helps them to recover quickly from the move. As a rule hardy and half-hardy plants raised in the manner described are planted

outside directly from the trays or pots after being hardened off. With true greenhouse plants the next move is usually from the trays singly into pots, or from small pots to larger ones. The compost for this purpose may be JIP₁, JIP₂ or JIP₃ (or equivalents), depending on the type of plant.

As already explained, cold frames often serve as an intermediate stage between the heated greenhouse and the open garden. Cold frames also serve for seed-raising and are particularly useful for starting such early vegetables as cauliflowers, onions and leeks. Many flower crops may also be raised by sowing seed in frames. Sowing may be done directly in the frame soil or in trays or pans placed in the frame. In unheated frames it is inadvisable to sow very early. Later sowing usually means better and quicker germination and more rapid growth of the seedlings.

Raising ferns from spores
Spores are botanically quite different from seeds but in the propagation of ferns they have a similar function (Fig. 25). Spores are borne on the undersides of fern leaves (usually called fronds) in special spore cases known as sporangia. When the sporangia begin to turn brown this indicates ripening, but to make quite sure of this those near the base of the frond should be examined with a hand lens. If these sporangia show signs of opening the fronds should be cut off and placed in large paper bags to dry.

Sowing of spores may be done at any time of the year but the sooner after harvesting the better. A suitable compost consists of 2 parts peat, 1 part loam and 1 part fine sand. All these should be passed through a 5 mm (¼ in) sieve. The sand and loam should be sterilized before mixing. Pots or pans and any crocks used should also be sterilized.

Cover the base of the pot with a layer of rough peat and fill up with the compost to within 1 cm (½ in) of the top. The containers should then be soaked with water, and after being allowed to drain the extremely fine spores are sown as thinly as possible without any compost covering.

After sowing, cover each pan with a piece of glass; stand the pans where they will receive moderate light conditions, but always ensuring protection from direct sunlight. For greenhouse ferns a temperature of 21°C (70°F) is required. Some gardeners keep the pans standing in saucers containing a solution of permanganate of potash

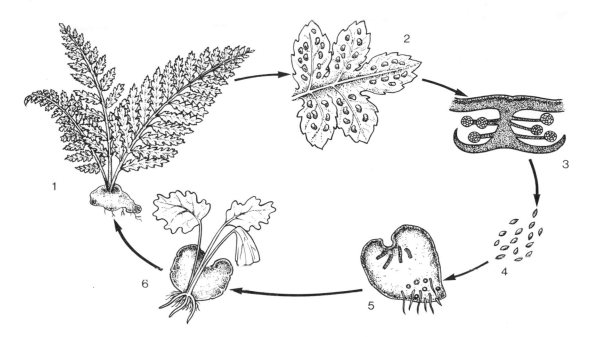

Fig. 25. Life cycle of fern. (*1*) Mature frond. (*2*) Underside of frond showing sporangia (*3*) in each of which is developed the spores. (*4*) Each spore is cap- able of developing, on germination, into a prothallus (*5*) from which, after a complex sexual phase, the mature fern develops.

which helps to prevent disease and also ensures that the compost does not dry out.

When the sporlings, usually termed 'prothalli', are large enough to handle they are pricked out in bunches consisting of three or four plantlets. Pans or trays filled with similar compost to that used for sowing are suitable. Alternatively a peat-based compost may be used. The little clumps are simply pressed gently on the soil sur- face without covering. No overhead watering is allowed until the little plants are well established and have produced fronds. Warm, fairly humid conditions are still maintained and when the young ferns are large enough they should be set singly in small pots.

8 Natural Vegetative Increase

Long before plants had evolved flowers, fruit and seed, many of them had developed a means of spreading and increasing by natural vegetative growths. The disadvantage of this method in nature is that the new plants are usually crowded together near their parents and there is, consequently, severe competition for air, light and plant nutrients. With garden species there is a need to lift plants that arise by this method and replant them properly spaced out. In other words, the gardener's role in natural plant increase is to divide or separate the plants that have arisen by such means as offsets, corms or bulbs.

Simple division

Simple division consists of splitting up individual plants into several smaller ones each with roots attached. Plants that lend themselves to this method are those with a tufted or matted habit.

Alpines provide many examples of this type including aubrieta, arabis, many dianthus, veronicas and violas. Division is a popular method of increasing herbaceous flowering plants, such as Michaelmas daisies, helianthus, peonies, rudbeckia, doronicum, pyrethrum and scabious. As a rule plants that flower in the spring should be divided in the autumn, while autumn-flowering subjects are dealt with in the spring. Certain plants, however, are liable to die in the winter if divided in autumn and should be transplanted in spring, or after flowering not later than August. These include pyrethrums, scabious, and German irises. Hardy lilies can be divided in spring or autumn. Several hardy herbaceous perennials, such as Michaelmas daisies, do not object to being transplanted during open weather in winter.

In dividing such plants as erigerons, the young portions around the outside should be selected and the older centre discarded. Some plants are difficult to tear apart, but may be prised into separate portions with a handfork. Quite a few can be chopped up neatly with a sharp spade, examples being achilleas and golden rod. This is a common method in nurseries.

Rooted portions of herbaceous perennials may be planted directly in the border or in a piece of ground used as a nursery bed. Here they can stay until a permanent site has been found for them.

Most are leafy plants and require plenty of moisture. The land on which they are planted should, therefore, be well and deeply cultivated, and farmyard manure or leafmould added to help in the retention of moisture. A balanced mixture of artificial foods may also be beneficial.

In nurseries most hardy herbaceous perennials are planted in wide 1.25 or 1.5 m (4 or 5 ft) beds. Spacing for most kinds is 30 cm (1 ft) apart each way, but some of the taller growing species, such as aconitums, *Chrysanthemum maximum* and delphiniums, are planted 60 cm by 45 cm (2 ft by 18 in). The plants should be made quite firm.

Strong-growing alpines may be divided in the spring or autumn and replanted in a nursery bed in the open or on the rock garden. Planting should usually be very firm, and protection from sun or drying winds is often an advantage. Water if necessary. Less robust alpines should be potted in suitable compost after division and may be

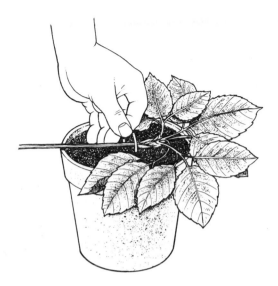

Fig. 26. Rooting a strawberry runner in a pot. Note the runner is pegged down to prevent movement of the plant in the compost.

stood in a cold frame and shaded from bright sunshine until they recover.

Some of the shrubs which may be increased by simple division are *Kerria japonica*, several of the berberis such as *B. stenophylla*, ericas and spiraeas. These may be lifted, divided and replanted in spring or autumn. Any shoot with roots attached will soon make a new plant. Some-

times a method called 'dropping' is used with shrubs intended for division. This consists of lifting the shrub and taking out some soil beneath it so that the plant is dropped 10 to 15 cm (4 to 6 in). This induces the stems to produce roots from their bases thus facilitating division.

New plants from old stems
Quite a number of plants produce stems that are capable of giving rise naturally to new plants. These stems have been modified or adapted for this particular function.

Runners These are slender stems which grow from the parent plant along the ground. The roots and shoots of a new plant arise at each node (leaf joint) and soon become established in the soil. The decay of the internode (the stem between the nodes) severs the connection with the parent plant. Sometimes the gardener assists the runners of strawberries to root by pegging them down (Fig. 26) or by placing a stone on the runner near each node.

Offset The offset is similar to the runner, and is found in the houseleek. In this case a slender stem grows out from the crown, bearing at its tip a fleshy rosette of leaves which is capable of taking root at its base.

Suckers The roots of raspberries spread in the soil a considerable distance from the parent cane. Often they produce buds at various points which develop into shoots coming above the ground to form new plants. These are called suckers and are

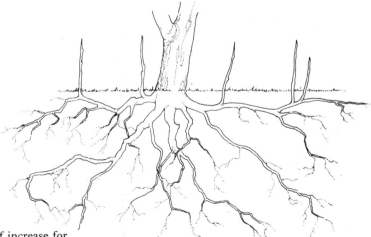

Fig. 27. Suckers form a ready means of increase for many woody perennials. They simply need to be severed from the parent plant together with a portion of root, and planted separately.

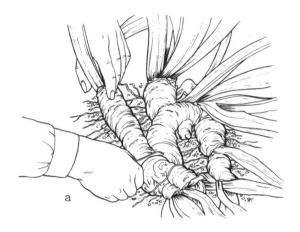

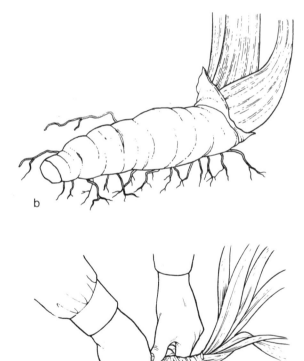

Fig. 28. Propagation by division – irises. (a) The current year's rhizotamous growth being separated from the old rhizomes. (b) The young, severed portion. (c) The new portion being positioned in the soil.

the principal means of increasing raspberries, apart from raising new varieties by seed. Other plants which produce suckers are plums, lilacs and filberts (Fig. 27).

Rhizomes The swollen structures of the Solomon's seal and the German iris which are found at the soil surface are also modified stems. Each with a bud attached will grow into a new plant. These swollen rhizomes are also storehouses of food for the young plant. Some plants, however, such as grasses and sedges, produce slender rhizomes like strawberry runners, but as they are underground they are often erroneously mistaken for roots. If such rhizomes are broken into pieces, each with one or more nodes, they may produce new plants (Fig 28).

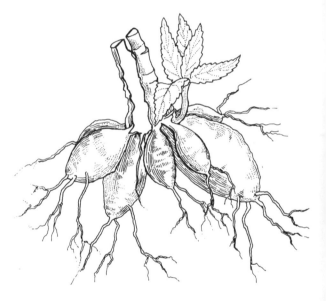

Fig. 29. Dahlias have tuberous roots, not true tubers. They can only be used for propagation if there is a piece of true stem attached having one or more buds.

Tubers Another type of underground stem is the tuber which is very convenient for reproduction. The potato is a common example and each of the buds or 'eyes' it possesses is capable of producing a shoot and roots. Sometimes tubers are cut in pieces and each piece having an eye may give rise to a plant. Other plants that produce tubers are Jerusalem artichokes and tuberous begonias. A clear distinction should be made between tubers and plants with tuberous roots, such as dahlias (Fig. 29). The latter are not tubers and can only be used for propagation if there is a piece of true stem attached having one or more nodes or eyes.

The tubers of such garden plants as potatoes, Jerusalem artichokes and begonias are stored over winter and planted in the spring. The latter keep well if bedded in dry sawdust or peat and kept in a frostproof store. Begonias are usually started into growth in heat.

Bulbs and corms

Some of our best garden plants are reproduced from bulbs and corms. These are also classified by botanists as modified stems. Corms are solid structures with one or more buds on the topmost side. When the corm is planted the buds grow upwards and produce foliage and flowers. In doing so the food contained in the corm is used up and the old corm shrivels. As the plants grow, however, a new corm is formed at the base of each shoot. Thus as many new corms may develop as there are buds on the old. Gladioli (Fig. 30) and crocus are good examples of corms.

In some species of this type (for example gladioli) large numbers of new buds arise on the old corm and develop into small corms about the size of peas. These 'cormlets' are often called spawn. They will grow into ordinary corms but take two or three years to reach flowering size. Spawn may be induced to arise by the artificial method of making two or three cuts across the base of the corm.

A bulb (Fig. 31) has a more complicated structure than a corm and can be observed by cutting a tulip bulb lengthwise. This is seen to consist of a short, thickened stem bearing roots on the underside and thick fleshy leaves on the upperside which encircle each other. Right in the centre and enclosed by the leaves is a large bud consisting of

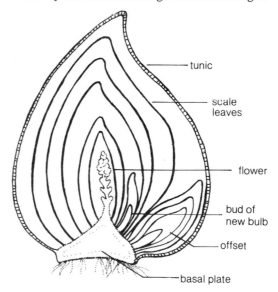

tunic

scale leaves

flower

bud of new bulb

offset

basal plate

Fig. 31. Longitudinal section of a typical tunicated bulb. The short, thickened stem bears roots on the underside and thick fleshy leaves on the upperside, which encircle the embryo flower.

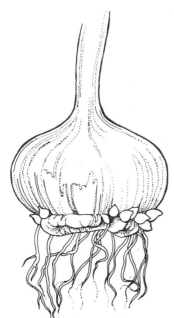

Fig. 30. Reproduction in corms. The cormlets are produced usually in large quantities round the base plate of the corm. The corm itself dies after flowering. The cormlets usually take two or three years to grow on into flowering-size plants. The example shown here is gladiolus.

the underdeveloped flower and foliage. Smaller buds may also be found on the short stem between the fleshy leaves. The latter serve merely as food storehouses and are protected externally by the scaly leaves.

Firm and compact bulbs such as tulips and narcissi whose scales encircle or nearly encircle one another are termed tunicated bulbs. Certain other bulbs, however, such as lilies, have scales that do not extend around one another in this way and are referred to as scaly bulbs. Plants with scaly bulbs are readily propagated by simply breaking off the scales and inserting them in a sandy compost.

Under suitable conditions the central bud grows upwards and produces flowers and foliage. Small buds, if present, also develop, but few of them produce flowers. All the buds give rise to new bulbs. A large bulb usually produces several daughter bulbs; a small bulb may produce one only, but this is normally larger than its parent. This sequence of events is repeated year after year so that the number of bulbs steadily increases.

Some bulbs reproduce themselves much more quickly than others. For tulips the rate of increase is appreciably greater than for hyacinths. The latter are so slow that artificial methods have to be used to stimulate the production of new bulbs. The method of doing so is described in Chapter 18.

Certain bulbous plants produce small or secondary bulbs called bulbils. For instance, on the flowering stalks of garlic and tree onions these are found instead of flowers. Each is capable of developing into an ordinary plant. Bulbils are also found on the stems of some lilies such as the tiger lily *Lilium tigrinum* and the golden-rayed lily of Japan *Lilium auratum*. With hyacinths, bulbils may develop on cut or broken bulb scales or may be artificially induced as shown in Fig. 32.

Bulbs succeed best in a deep loam well supplied with decaying animal and vegetable matter. They dislike fresh manure and are best grown in land that has been well manured for a previous crop. Good deep cultivation is essential and the land should be well drained.

Tulips, gladioli and irises like lime in the soil and it is often an advantage to apply some finely ground limestone before planting. Narcissi are not so partial to lime, while certain bulbs, such as some lilies, most definitely dislike lime.

For large scale production in fields bulbs are usually machine-planted, but hand-planting is used for smaller areas. In the case of bulbs such as tulips and narcissi a popular method consists of 1.25 m (4 ft) wide beds. The bulbs are planted in rows about 23 cm (9 in) apart across the beds and the bulbs are spaced to one or two times the diameter of each. Depth of planting varies according to the species but tulips and narcissi are usually planted 8 to 10 cm (3 to 4 in) to their base.

After planting, weeds can be kept in check by light surface cultivation, hand weeding or the use of a pre-emergence herbicide such as Paraquat, which can be safely used until shortly before shoot emergence. Having destroyed the weeds present, a residual soil herbicide such as CIPC/ Diuron/IPC in a proprietory mixture can be applied. Such herbicides may keep the beds free from weeds for several months.

Fig. 32. Spawning hyacinth. Two methods are used for inducing hyacinths to form bulbils. The first, shown left, is to make a clean cut horizontally across the basal plate of the hyacinth bulb and then to make three vertical cuts in the shape of a star. The alternative is to scoop the whole of the base plate out, leaving a conical cavity.

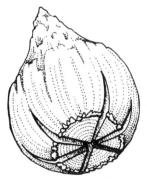

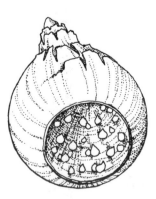

Further routine operations in bulb production, especially with tulips, consist in the periodic removal of diseased plants when they can be seen (see chapter on pests and diseases). When the bulbs come into flower those not true to type should be pulled up and all weak or diseased-looking specimens should also be removed. Pulling off the flowers immediately they are seen to be true to variety is a common practice. This is done to give the bulbs a better chance to develop as the blooms take nourishment from the plants. De-blossoming is regarded as essential for hyacinths, tulips and irises, and it is also beneficial in the cases of gladioli and narcissi.

When bulb production is the primary aim it is usual to lift the bulbs annually when the leaves have died down. A period of dry sunny weather is a great advantage to ripen the bulbs properly. Some bulbs such as tulips are lifted as early as June, but late gladioli may not be ready for lifting until October and should be left as long as possible.

Lifting should be done with a fork and the bulbs should be carefully collected and allowed to dry, preferably in an open shed. When dried the bulbs are cleaned, separated and freed from rubbish. Finally, they are sorted into sizes.

9
Artificial Methods of Vegetative Increase

Any portion of a plant, whether from the root, the stems or the leaf, when separated from the parent plant for propagation purposes, is usually described as a cutting.

Some plants and trees such as poplar and willow have root beginnings or 'initials' already present in their stems. With these it is only necessary to bring the stems into contact with moist soil or water to arouse their sleeping roots into activity and growth.

Such plants, however, are exceptional and with the majority of stem cuttings new roots have to be initiated. Botanists tell us that an incitement for root production is provided when the stem is severed from its parent. This is called wound shock. The first reaction is the formation of a tissue, called callus, which protects the wound in the same manner as a scab protects a cut on one's finger. Next, roots are produced with the apparent object of keeping the isolated piece of stem alive.

Both the callus tissue and the roots arise from the cambium, that layer of actively dividing cells found beneath the bark and mentioned in Chapter 1. It is interesting to note that, although there is no fundamental difference between cambium in the roots or stems of plants, in the roots it may give rise to new shoots and in the stems to roots.

The formation of callus is a necessary preliminary to rooting, although roots do not arise in the callus but from the cambium tissue immediately behind it, and quite frequently they appear also further up the stem. In budding and grafting it is the cambium which enables a union to be formed between the parts pressed together.

The type of cutting, including its degree of maturity and the time and manner of securing it, affect propagation, but these factors will be discussed in succeeding chapters. As a general rule cuttings made from young growths of the current year are preferable. They should always be secured with a sharp knife so as to give a clean cut which promotes rapid healing. Jagged wounds and mangled tissue may cause decay before rooting can take place.

Conditions that promote rooting

The essential conditions for rooting cuttings are the presence of moisture, air, and a suitable temperature. The need for moisture is obvious, seeing that the cutting has been isolated from its former source of supply and is in real danger of being desiccated. This applies particularly to leafy cuttings as their leaves are still giving off or transpiring moisture. Early insertion, therefore, in a medium such as moist soil, sand or compost is essential. Moreover, in the case of leafy cuttings loss of moisture must be reduced to a minimum. This is usually achieved by keeping the cuttings in a closed and shaded frame, and maintaining a humid atmosphere.

The formation of callus and the production of roots at the base of a cutting calls for increased cambial activity and consequently accelerated respiration from the tissues. This indicates the necessity for good aeration around the base of the cutting, and is the reason why sharp sand is often used in striking cuttings. Sand allows air to enter freely. When sand and peat are mixed together this forms an excellent rooting medium because

66

while the sand promotes aeration the peat retains moisture.

Rooting of cuttings is usually forwarded by heat, but naturally this varies according to the type of plant. Thus, cuttings from stove plants, such as crotons, require a higher temperature than chrysanthemums or dahlias. An important rule, however, is that the air temperature should be below that of the medium in which the cuttings are inserted.

This means that while callus production and rooting are forwarded, the development of the buds and leaves is not so rapid. Experience has proved that rooting should preferably precede the growth of the shoots. This is why bottom heat is often used in the propagation of cuttings.

Another general rule is that soft leafy cuttings respond to higher temperatures than their mature counterparts. Hardwood cuttings usually take a considerable period for callusing and rooting whatever the conditions, so that heat may result only in the immediate growth of the shoots and leaves. This sometimes occurs when hardwood cuttings are planted outside in spring, and as the cuttings have no roots to make good the loss of moisture from their leaves they may shrivel up and die.

When leafy cuttings are inserted in a frame they usually require shade from bright light to prevent wilting. If shaded continuously, however, the leaves are unable to manufacture food (light is essential for this process) and the cuttings may die of starvation. The correct procedure in such circumstances is to shade for a few days and then gradually reduce the shade until they are fully exposed. Hardwood cuttings do not require light until they start to produce shoots.

The conditions that are suitable for rooting leafy cuttings, that is, warmth and moisture, are also very favourable to certain plant diseases. Strict hygiene is, therefore, necessary in propagating frames. Although it is necessary to retain as much foliage as possible on the cutting, the leaves that are likely to touch the soil should be cut off. The frame should also be examined frequently and any dead leaves carefully removed.

Effect of auxins or hormones

Botanists have shown that all growth and development in plants is controlled mainly by highly active chemicals called 'hormones' or 'auxins'. These substances are produced chiefly at such growing points as stem tips and in the buds and leaves. From here they are conducted to other parts of the plant where they exercise control over the centres that initiate growth and affect the type of growth produced. For instance, when a shoot bends towards the light this is due to the growth-stimulating effect of a hormone on the shaded side of the stem.

Hormones are known to stimulate the production of roots in stem cuttings, and a cutting deprived of buds or leaves, and consequently its ability to produce hormones, may fail to root because of this. Plant hormones have been the subject of considerable research and the more important ones have been identified. Moreover, certain substances have been discovered which have a similar effect on plant growth. These are usually called 'growth regulating substances' and can be produced synthetically.

Auxins are now widely used as an aid in the rooting of cuttings by both nurserymen and amateur gardeners (Fig. 33). Their principal effect is to accelerate the production of roots—cuttings so treated will usually produce more roots than when the chemical is not used. This means that plants raised from treated

Fig. 33. Encouraging root growth in cuttings by 'hormone' rooting powder. (For hardwood cuttings, removing the rind at the base for, say, 1.2 cm (½ in) may speed up rooting.) Moisten the base of the shoot and dip the cutting into the powder. Shake to remove excess powder before inserting into compost/soil.

cuttings can often be planted out earlier and will usually grow faster than plants from untreated cuttings.

On the other hand, hormones cannot induce rooting where cuttings are taken from plants which have never been known to produce roots by this method. Otherwise, these chemicals are used for both hardwood and softwood cuttings. Hormones probably promote rooting in temperatures down to 10°C(50°F) but are much more effective when the soil temperature is at least 16°C (60°F).

For root stimulation in cuttings three different chemicals are now used as follows:

1. Indolyl acetic acid (IAA).
2. Indolyl butyric acid (IBA).
3. Naphthalene acetic acid (NAA).

IAA was the first auxin to be discovered and is found in plants while the other two are not known to occur in plants. For general use, IAA and IBA are preferred to NAA because although the latter does stimulate rooting it is known to retard shoot growth in some cases.

Auxins are available either in liquid or powder form. Dusts or powders contain the active ingredients in an inert carrier such as talc, kaolin or fine charcoal. Proprietary dusts usually contain from 0.1% to 1.0% by weight of IBA or IAA. Some firms provide three different strengths such as No. 1 for softwood cuttings, No. 2 for semi-hardwood and No. 3 for hardwood cuttings. Obviously the highest concentrations are used for the hardwood cuttings.

Hormones in solution are widely used on commercial nurseries as they allow convenient treatment of large numbers of cuttings. One method is to stand hardwood cuttings with their bases in an IBA solution with a concentration of 50 to 200 parts per million of the active ingredient for 18 to 24 hours. Weaker strengths are used for softwood cuttings. Another method consists of giving a quick dip into a concentrated solution containing 50% alcohol with the hormone at about 25 times the normal strength. This method is sometimes used for difficult subjects such as rhododendrons and apple stocks.

For hormone treatment of cuttings by the amateur it is generally advisable to use a powder. Care should be taken to ensure that the powder is of the correct strength for the type of cutting and to read the maker's instructions. As a rule the base of each cutting is moistened before inserting in the powder. Avoid coating more than the base of each cutting with powder, and tap off any excess on the rim of the container.

Hormones can also be used to treat wounds or cuts made during the process of layering, which is described in Chapter 13. There is little point in giving a list of species for which hormones can be used as the range is now so comprehensive.

10
Mist Propagation

The rooting of softwood leafy cuttings under an intermittent spray or mist of water is a technique now widely used by nurserymen and by some amateurs. The aim of misting is to maintain continuously a film of water on the leaves, thus reducing transpiration and keeping the cuttings turgid until rooting can take place. In this way leafy cuttings can be fully exposed to light and air because humidity remains high and prevents damage even from bright sunshine. In practice misting accelerates rooting and promotes the striking of difficult subjects such as hollies and rhododendrons. Mist will also prevent disease in cuttings by washing off fungus spores before they can attack the tissues.

Way back in the mid-fifties the phrase 'Hot bottoms—misty middles—and cold tops' was used to introduce this new technique to the enthusiastic amateur and commercial grower. Since then many hundreds of keen gardeners have installed 'mist' in one form or another, and in most cases they say it is the best thing they have ever spent money on. Many of these old units are still working after fifteen years use, so the actual cost of 'mist' is quite low when one considers the vast number of rooted cuttings that one can produce with a one-jet unit. The choice of equipment is quite large, so 'shop around' and get as much information from your friends and the manufacturers as possible before deciding on the type to invest in. For instance, there are varying makes of 'transistorized units', 'silicon chip units', 'solar' and others to choose from. Then there is the sensitive 'time clock' being used by a few commercial growers and needing specialist set-

ting. Remember that no electrical equipment lasts forever so a look at the future service prospects is important.

While the leaves in this process must be kept continuously moist it is important that only a minimum of water should be used. This is because excessive water leaches out nutrients from the compost and may cause starvation. Moreover, a directly injurious effect on the cuttings can occur from over-watering. Hence, it is essential to use nozzles capable of producing a very fine mist.

There are three methods of control. One is by an electric time-clock mechanism. Thus, the time-clock may be set to give a spray burst every three, every five or every seven minutes. The disadvantage of this method is that the intervals between misting periods remain the same regardless of weather conditions. Thus, if the time clock was set to give a spray burst every five minutes this may be excessive under dull weather conditions, but might be inadequate during bright sunny periods.

The second method of spray control consists of a sensing element sometimes called an electronic leaf. This is used in conjunction with a solenoid valve and a switching controller installed in the water lines. The electronic leaf is placed among the cuttings and dries out or loses superficial water at exactly the same rate as they do. At a certain stage of dryness the sensing element activates and opens the solenoid valve so that water is supplied to the nozzles. When moistened by the spray the electronic leaf causes the solenoid valve to close again and misting is suspended.

In this way the frequency of spraying periods is controlled in relation to temperature and light intensity. Thus, on a bright day misting may occur every 2 to 3 minutes, but on a dull day at not less than say 15 minute intervals and only once or twice during the night.

The third method of control is perhaps the most efficient. It utilizes a light-sensitive control box which is wired into the system near the bench but out of range of the spray. This solar controller switches the water on and off depending on the brightness and duration of sunlight, which will be in proportion to the evaporation of moisture from the cuttings.

The solar controller is virtually maintenance-free, for unlike electronic leaves it does not need to be cleared of algae and hard water deposits from time to time.

Benches for mist propagation are usually 1 m (3 to 3½ ft) wide and the mist propagation unit (Fig. 34) may consist of one or more nozzles. The nozzles are usually spaced 1 to 1.25 m (3 to 4 ft) apart and 38 cm (15 in) above the bed.

The period from May until September is the ideal time to root cuttings under mist. During this period a wide range of softwood cuttings will be available from shrubs and other plants and can be rooted without the aid of bottom heat. Under-bed heating, however, is certainly an advantage with mist even in summer and gives quicker root-

Fig. 34. A mist propagation unit showing heat and water controls. This type of propagating equipment will hasten the rooting of particularly difficult subjects.

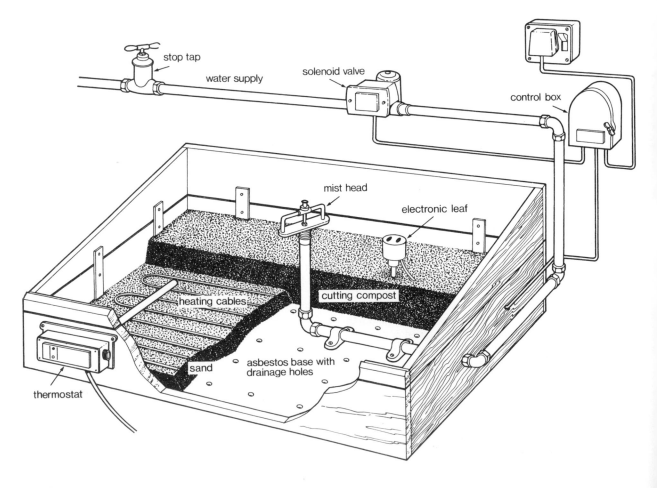

Table 10.1. Rooting results from mist propagation

Species	Variety	Type of Cutting	Date	Hormone No.*	%	Approx. Rooting Time (days)
Trees & Shrubs						
ABELIA	All	Soft	June–Aug	1	100	21–30
ACER	Some only	Just firming	July–Aug	2 or 3	Low	60–80
ARBUTUS	All	Medium	Oct–Nov	2 or 3	100	20–30
AUCUBA	All	Any	Jun–Feb	1 or 2	100	20–30
AZALEA	All evergreen	Just firming	July–Oct	1 or 2	100	30–40
AZALEA	mollis	Soft	June–July	2	Low	50–80
BUDDLEIA	All	Soft	June–Sept	1 or 2	100	15–20
BUXUS	All	Any	Oct–Nov	1 or 2	100	20–25
CALLUNA	All	Just firming	Sept–Nov	1 or 2	95	30–50
CAMELLIA	Most	Firm	Oct–Nov	2	100	50–80
CHOISYA	All	Firm	Sept–Nov	2	100	20–30
CISTUS	All	Soft or hard	Sept–Dec	1 or 2	100	25–40
CLEMATIS	Some–try all	Soft internodal	Apl–June	2	Varies	30–60
COTONEASTER	All	Any	June–Nov	2	100	20–30
CYTISUS	Most	Just firming	July–Sept	1 or 2	80	40–50
DAPHNE	Some–try all	Firm	Oct–Feb	2	Varies	50–80
DEUTZIA	All	Soft to firm	July–Aug	1 or 2	100	20–30
ELAEAGNUS	Most	Firm	Aug–Nov	2	Varies	36–65
ESCALLONIA	All	Any	June–Sept	2	100	20–30
FORSYTHIA	All	Soft	June–Aug	1 or 2	100	20–30
FUCHSIA	All	Soft shoots	Any time	Nil	100	7–15
HEBE	Most	Soft to firm	July–Nov	1 or 2	90	15–20
LEPTOSPERMUM	Most	Just firming	July–Aug	2	75	20–30
MAGNOLIA	Some–try all	Medium soft	June–Aug	2 or 3	Varies	40–60
PHILADELPHUS	All	Soft	July–Aug	2	100	25–40
POTENTILLA	All	Soft to firm	June–Sept	1	100	20–30
RHODODENDRON	Some–small-leaved	Firming	Sept–Oct	2 or 3	Varies	60–90
VIBURNUM	Most	Medium soft	June–Oct	2	80	30–60
Conifers	Most are good	Firming–take with heel	Sept–Oct	2	Varies	40–100
Herbaceous Plants	Lupins, phlox, asters, delphiniums & many others	Soft tips	As soon as available	1	100	10–30
Rock Garden Plants	Helianthemum, aubrieta, phlox, dianthus, etc.	Soft to medium	July–Oct	1	75–100	10–30
House Plants	Many good	Soft to firm	Try any time	1 or 2	Varies	10–?
Fruit						
STRAWBERRIES	All excellent	Take runners before they touch the ground to avoid redcore disease				10–20
LOGANBERRIES } BLACKBERRIES }	Both root easily from young tips					

Follow manufacturer's instructions for strength of hormone powder to be used.

ing. Heat also allows considerable extension of seasonal propagation which for some plants may be practically the whole year round.

Installation is so important that only a qualified electrician should be employed for this not too difficult but vital job. Normally, the heating cables are laid on sharp grit or washed sand and then covered to a depth of 5 cm (2 in) with the same material. The recommended loading for mist propagation is 15 watts per 30 cm sq (per sq ft). A thermostat should be installed and the bed temperature maintained at 18 to 24°C. (65 to 75°F). Temperatures may be raised very slightly according to time of year and the subject being rooted.

Trays or pots filled with the rooting medium may be stood on the sand. Benches must be strong enough to carry extra weight of water and heavy sand. Drainage is critical. The makers give detailed instructions on the method required. Alternatively, the cuttings may be inserted into the medium which is placed over the heating wires to a depth of about 10 cm (4 in). The rooting medium may consist of sharp washed sand only, or of a mixture of sharp sand and peat. Perfect drainage is so necessary; a clean washed coarse grit is usually obtainable locally. Some propagators advise that for mist propagation peat should be used sparingly, if at all, as it tends to impede drainage. Boxes (approx. 7 to 8 cm (3 in) deep) with plenty of drainage holes are being used more and more for ease of removal after the cuttings are rooted.

Plenty of ventilation is needed on most days. Dead or decaying foliage should be removed quickly. Algae growth can be quite a problem in some areas. A new spray is now on the market and is proving very successful. The 'leaf' should be cleaned by rubbing the surface on to a flat piece of '0' sandpaper with a circular movement, or one of the denture cleaners obtainable from a chemist may be used. Be sure to wash the surface well with clean water after treatment.

The taking and preparation of cuttings for mist propagation is normal except that with some aspects there is more latitude. Thus, rather soft unripe cuttings have been struck under mist. At the other extreme cuttings that have gone a little woody and which would be slow to root in a closed propagating case, will probably strike readily under mist. Again, longer cuttings can be rooted with the aid of mist than can be rooted without it. Experience shows too that rather less care is necessary in the preparation of cuttings, for internodal cuttings from many shrubs root just as well under mist as those carefully cut just below the node. Hormone treatment is recommended, for even under mist it results in accelerated rooting. Many plant cuttings respond well to treatment. Follow the makers' recommendations as to strength, but the addition of 50% of a fungicide mixed with the hormone also helps; 'trial and error' seems the answer.

When the cuttings are rooted misting should not cease abruptly as this may cause drying out of the young plants followed by scorching. Instead, a weaning-off process should be adopted in which misting is continued but the number of sprays per day gradually reduced. Where the cuttings have been planted in boxes or pots these may be removed to another greenhouse or frame and the weaning-off process completed there. Alternatively, the rooted cuttings may be potted, removed to a shaded house and sprayed over occasionally until established. Whatever method is adopted it is obviously an advantage to clear the mist cutting beds as soon as possible if these are required for further propagation.

A very wide range of plants (Table 10.1) may be mist propagated and although one may have some failures with difficult subjects it is not often that all the cuttings in any given batch refuse to produce roots. The rooting of conifers, one of the slowest groups to root, is greatly accelerated by mist. Also, it should be noted that a vast range of seeds can be germinated under mist, but great care must be taken with 'weaning-off'; shade for a few days.

11
New Plants from Cuttings

Cuttings are made from the stems, leaves and roots of different plants and the type of cutting used depends on the plant being propagated, the time of year, the facilities to hand and previous experience.

Stem cuttings

Next to seed, stem cuttings are the most convenient and popular method of propagation. There are, however, a few general considerations which help in the selection of suitable cuttings.

First of all it is essential for the cutting to have a sufficient reserve of food stored within its tissues to enable it to remain alive until roots and shoots have been produced, when it can secure and manufacture food for itself. The foods called carbohydrates, that is sugar and starch, are the most important and it is found that shoots with a high carbohydrate content in relation to nitrogen give the best results.

As a general rule, cuttings from young plants root best, but if older plants are cut back hard they can often be induced to produce suitable material. An alternative to inserting the stem cuttings into a cuttings compost is to use moist sphagnum moss and to wrap the cuttings tightly in a polythene wrap to prevent evaporation (Plate 2).

Broadly speaking there are three types of stem cuttings, namely hardwood, softwood and semi-hardwood or half-ripe.

Hardwood cuttings

Hardwood cuttings are made from the mature stems of shrubs and trees, and provide the simplest of all methods of propagation. Such cuttings are often easy to secure and root readily in ordinary soil in the open. Hardwood cuttings are usually made from firm stout stems of the current year's growths, but some plants (tamarix) grow freely from stems 2 to 3 years old. The length of these cuttings varies from about 10 to 30 cm (4 to 12 in), but cuttings of such plants as mulberry and willow may be as much as 1 to 1.25 m (3 to 4 ft) long.

It is often advantageous to take hardwood cuttings with a 'heel', that is, with a piece of older wood attached to the base (Fig. 35a). This is essential in the case of conifers which are propagated from small sideshoots. Cuttings secured without heels are usually cut just below a leaf joint or node (Fig. 35b), but in the case of clematis, internodal cuttings with one pair of leaves at the top are preferred. For plants with pithy stems it is advisable to sever the cutting just at the point where the current year's growth originates, as here the pith area is reduced to a minimum. On all cuttings at least one bud near the tip is necessary for extension growth. Buds are not required near the base as roots are produced not from buds, but from the inner cambium tissue.

Hardwood stem cuttings will often root better if their bases are etiolated or blanched before being removed from the 'parent' plant. This is achieved by cutting back the stock plant to within a few inches of ground level. The young shoots that arise at this point are then kept earthed up with soil as they grow. Certain plants for fruit stocks when treated in this way can be propagated

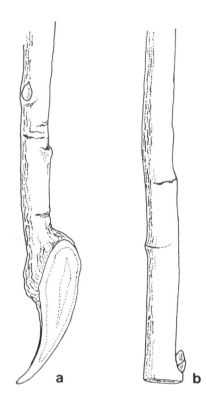

Fig. 35. Hardwood cuttings. A 'heel' cutting (*a*) and a nodal cutting (*b*).

from cuttings (see p. 101).

Autumn is probably the best time to secure hardwood cuttings. They may then be planted immediately or tied in bundles and laid in moist soil until the spring. In either case callusing may begin at once, which allows for quicker rooting in the spring. A wide range of deciduous plants is increased by hardwood cuttings, including the soft fruits, currants and gooseberries, and many shrubs and trees, such as poplars, privet, red-stemmed dogwood, tamarix, flowering currants and forsythia. Some evergreens are also increased in this way, for example, common and Portugal laurels, *Lonicera nitida* and *Aucuba japonica*. Conifers of the chamaecyparis type also root from heeled cuttings, but are usually planted in a cold frame.

Heated bins A different technique with hardwood stem cuttings was developed by Dr. B. H. Howard of the East Malling Research Station.

This was designed to facilitate the propagation of apple, pear and plum fruit tree root stocks from hardwood cuttings rather than by the more laborious method of layering. For this purpose vigorous young shoots 60 cm (24 in) long are used, these being secured from special plantings of the root stocks. Plum stocks and the quince which is used as a rootstock for pears which start into growth relatively early in the spring are taken in the autumn, whereas the seasonally later developing apples are taken in February or March.

The success of this method depends on the use of heated bins to secure quick rooting. In these bins bottom heat is effected by thermostatically controlled electrical warming wires laid on a well-drained insulated floor, the loading being 15 watts per 30 cm sq (per sq ft). The rooting medium consists of equal parts grit and coarse sphagnum moss peat laid over the wires to a depth of about 25 cm (10 in).

After hormone treatment the cuttings are inserted thickly—250 per 30 cm sq (per sq ft)—with their bases about 2.5 cm (1 in) above the wires. Subsequently, a temperature of 21°C (70°F) is maintained. One month later the cuttings should be fully rooted and be ready to transplant.

Hardwood cuttings in the open Many cuttings of the hardwood type can be rooted easily in the open garden. Select a sheltered position and soil which is free from weeds. The cuttings are planted simply by putting down a line and taking out a shallow trench having a straight vertical side next to the line. Place the cuttings in the trench a few inches apart and at least three-quarters their length deep. The soil is then replaced and trodden in firmly against the stems (Fig. 36). If there is more than one row it is usual to allow at least 30 cm (12 in) between them. As previously explained, hardwood cuttings may be planted either in autumn or in spring. Examples of species rooted in this way include blackcurrants, gooseberries, forsythia, ribes, poplar, willow and many evergreens such as common laurel, *Lonicera nitida*, *Viburnum tinus*, hebes and griselinia. Conifer cuttings may root in warm sandy soil in a sheltered position.

Softwood cuttings

A very wide range of plants are raised from softwood cuttings. These have the advantage of being

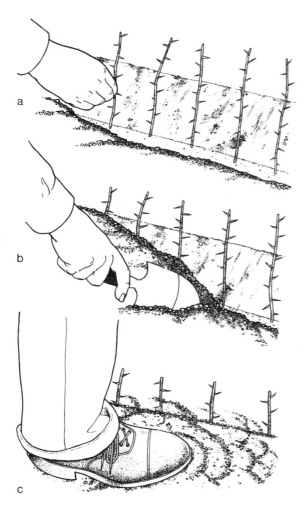

autumn. In nurseries violas are often de-blossomed with a knife so as to induce them to produce suitable shoots for propagation.

It is estimated that over 2,000 alpines may be increased by cuttings. These are secured from various plants from spring until the autumn. Frequently cuttings are taken prior to flowering; often they may be had after flowering. Sometimes it is a good idea to cut back plants after flowering to induce the production of cuttings. Aubrietas, alyssum and arabis are examples of plants treated in this way. The important point is to secure young, firm growths whatever the season.

Ornamental shrubs in great variety are readily increased from softwood cuttings, that is cuttings made during the summer from the immature tips of current growths. Sometimes shrubs of certain species are grown in pots so that they can be taken into heated greenhouses and forced into growth out of season to produce shoots for use as soft-wood cuttings. Softwood cuttings from shrubs should be reasonably firm and if the shoots snap off cleanly without bending this indicates that they are not too old for propagation.

In preparing softwood cuttings it is usually recommended to cut below a node, but quite a number of plants can be rooted from internodal cuttings (Fig. 37). These include verbenas, anti-rrhinums, chrysanthemums, *Hydrangea macro-phylla*, fuchsias, heliotrope, calceolaria, penstemons, dahlias, violas, lavenders and clematis.

The principal advantage of internodal cuttings

Fig. 36. Hardwood cuttings of gooseberries. A trench is made by inserting a spade into the ground vertically and levering the spade away from the vertical edge. The cuttings are inserted vertically against the vertical edge (*a*), the earth pushed back against them (*b*) and firmed with the foot (*c*).

Fig. 37. Inter-nodal cuttings. The diagram shows how an internodal cutting is taken and planted.

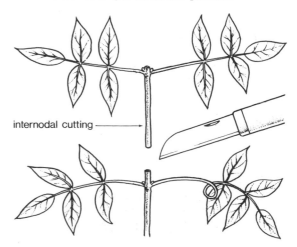

internodal cutting ⟶

quick rooting, but they must always be given the protection of some form of glass covering. A great many herbaceous perennials, such as delphiniums, lupins, phlox, dahlias and chrysanthemums, are readily increased in this way. The young shoots produced in the spring are mostly used, each providing a basal cutting. Other miscellaneous plants of a herbaceous type, violas, penstemons and pelargoniums, provide cuttings that may be rooted during the summer and

Fig. 38. Softwood cuttings of geraniums. Note where the cuts are made and which buds and leaves are removed. Set the cuttings around the edge of a pot of cuttings' compost to root.

occurs in circumstances where propagating material is scarce. For instance, in an effort to increase the number of young plants which are still small one may wish to use their tips as cuttings. By taking an internodal cutting this allows another node to be left on the young plant. Sometimes it is difficult to find the right type of softwood cutting on a plant. But it may be possible to induce suitable growths by cutting back a shoot or pinching out its top.

Some plants have pithy or hollow stems and these are best propagated by nodal cuttings. Other plants of similar type may have short laterals which can be conveniently taken with a 'heel'. Softwood cuttings to avoid are thick, coarse, over-vigorous shoots; flowering shoots are also inadvisable but are sometimes used when there is nothing else available. In this case the flowering tip should be cut off.

Softwood cuttings (Fig. 38) are usually prepared 5 to 10 cm (2 to 4 in) long. Any of the lower leaves likely to touch the propagating medium after insertion should be removed. Some trimming of the tips and shortening large leaves is often advisable to prevent wilting, but it is important to leave as much foliage as possible, as this is known to promote rooting.

A special type of softwood cutting used in the propagation of pinks is called a 'piping'. Pipings are secured simply by pulling off the ends of young shoots (Fig. 39). The lower leaves are removed and the cutting is ready for insertion. As a rule softwood cuttings are inserted immediately they are prepared, but a few species, such as the cactus group (Fig. 40) and zonal pelargoniums, may be laid on the bench to dry before planting. This is not essential and experience will show whether or not it is worthwhile.

Unless mist propagation is used (see Chapter 10) practicaly all softwood cuttings must be given the protection of a glass covering or a plastic shroud to allow the maintenance of a moist atmosphere which retards the loss of moisture from the cuttings. Rooting is also promoted by warmth particularly when it is applied directly as bottom heat to the propagating medium. In such

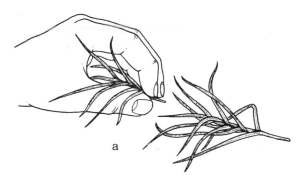

Fig. 39. Carnation cuttings may be taken as 'pipings'. These are simply pulled from the parent plant, the parent plant being gripped between thumb and finger just below the node from which the piping is to come (*a*). Set the pipings around the edge of a pot as shown (*b*).

longer than in heated frames, but this may not be of great importance to the amateur. The position of the frame is important. If it is sited in full sun it will, of course, be warmer than a shaded frame, and this results in quicker rooting, but unless precautions are taken the cuttings may be scorched and killed if fully exposed to strong sun.

Fig. 40. Cactus cuttings. Many species of cacti are readily propagated by simply severing a section from the parent plant (*a*), allowing it to callus for several days and then setting the cutting in a very sandy compost, preferably with bottom heat (*b*).

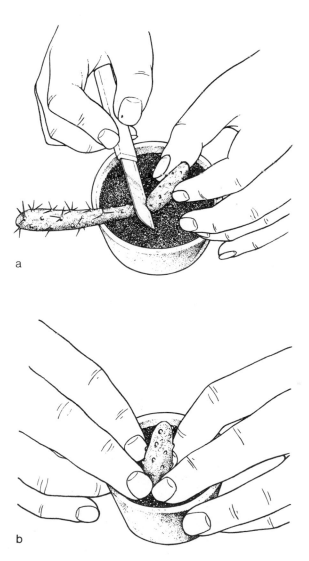

circumstances softwood cuttings have been known to root in 14 days. The only type of cuttings likely to fail are those with soft woolly leaves such as several alpines which quickly rot in the moist, warm atmosphere.

Certain shrubs (cistus and azaleas) root better under conditions that are less close, such as the open staging of a greenhouse. This method is also used for the propagation of chrysanthemums and dahlias which are either inserted in trays or pots, or in beds of sand or compost laid on the staging. In such circumstances some shading may be necessary for a few days and is best provided with sheets of paper laid lightly over the cuttings.

Cold frames in the open garden are suitable either for softwood cuttings or those made from semi-mature wood. Rooting, however, takes

A cold frame provides the amateur with the means of propagating a wide range of herbaceous perennials and throughout the summer and autumn it is convenient for the increase of many alpines and shrubs.

The propagating medium

As a general rule when a cutting has been prepared it should be inserted in a moist medium as soon as possible. This applies in particular to softwood leafy cuttings which lose water rapidly after they have been isolated.

Various mixtures and composts are used as propagating media, but sharp sand is the basis of most of them. Sometimes sand is used alone, and although it is a very good rooting medium, cuttings must be transferred from it into soil immediately they have rooted, as sand contains no plant foods.

Perhaps there is nothing better than a mixture of peat and sand. Suitable proportions are 2 parts sand and 1 part peat by bulk. Such a compost, while being well aerated, is also retentive of moisture. Plants belonging to the heath family do particuarly well in a peaty compost, but practically all cuttings may be rooted in it.

Perlite is an alternative to sand, and used alone it is recommended for different subjects. It is expensive but may be used over and over again. Another lightweight, yet absorbent medium is vermiculite. This is a form of mica found in the USA and in South Africa, and is a light flaky substance with a great capacity for holding water. It often remains moist in a cutting bed for periods of 14 days or even longer without watering.

Considerable success has been claimed for a mixture of sand and live chopped sphagnum moss. The compost is prepared by chopping up the moss very finely. As much silver sand as the moss will hold is then mixed with it. When kept moist this is an excellent medium for difficult cuttings.

There are various other mixtures used to root cuttings. Different proportions of sand, peat and soil, such as 3 parts loam, 1 sand and 1 peat (all by bulk) are sometimes recommended. A common practice is to finish off the surface of the soil or compost with a sprinkling of silver sand. This means that when the cuttings are being inserted, some of the sand on the surface trickles to the bottom of the hole where it is in direct contact with the base of the cutting.

Many cuttings are tricky to establish when rooted and stand a better chance of survival if dibbed into individual units as soon as they are taken. Compressed peat pellets which expand when stood in shallow trays of water can be used with a wide range of subjects. They are known as Jiffy 7's and the peat is held together by a thin plastic net that will apparently allow roots to grow through it. Many nurserymen use these pellets to root geraniums and chrysanthemums, and a wide range of shrubs and tender plants will root well in the peaty mixture.

On a smaller scale the amateur can also make use of 'rooting bags'. These miniature versions of the growing bag contain a compost suitable for inducing roots on cuttings. Slits are made in the bag, water poured in, the cuttings inserted and then left to root. When they are established the plants can be potted individually.

As a rule a special compost, pure sand, or a substitute for sand such as perlite, is the propagating medium used in heated frames. For cold frames ordinary soil lightened by a little pure sand is quite suitable.

In frames cuttings are either inserted directly into the prepared soil or compost in the frame or into compost contained in seed trays or pots, which are placed in the frame. There is a lot to be said for the latter method. It allows cuttings to be transferred from the frames at any time, perhaps for hardening off after rooting, without transplanting them. Pots and pans are ideal for small quantities of cuttings. Moreover, it is well known that difficult species often root best when inserted around the pot sides. No doubt this is due to sharp drainage.

There are several different types of heated propagators commercially available; Fig 41a shows one of them. It is quite possible to make your own heated propagator (Fig 41b and Fig. 42), provided you either have expert electrical knowledge or have access to a qualified electrical contractor. It is essential to ensure that the heating system is absolutely safe.

Whether the compost is contained in receptacles or in frames, make it reasonably firm, and ensure that it is nicely moist when the cuttings are being planted. Insert the cuttings with a small dibber in rows a few inches apart so that they almost touch one another, and make each firm. Afterwards give a good watering to settle the compost around the cuttings.

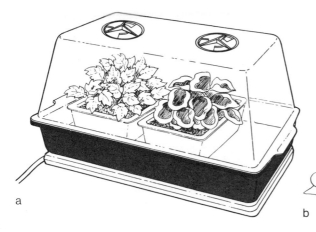

a

b

Fig. 41. Propagators. (*a*) A single, heated propagator with seed tray and cover. (*b*) Home-made propagator with wooden sides, rod thermostat and soil-warming cables. A polythene shroud is used as a cover.

Fig. 42. Home-made propagating frame consisting of eight glass sheets held in position by a wooden framework. Each pane can be removed separately. The frame has a slatted floor over the heating area which is warmed by a tubular heater.

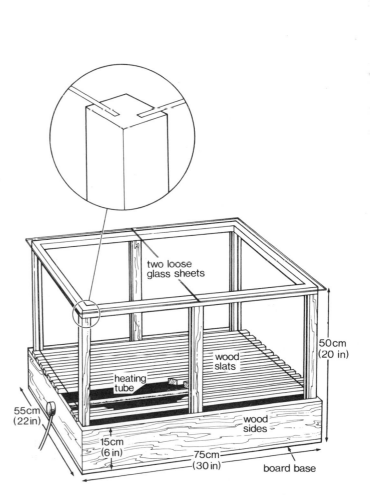

The after-care of cuttings in frames consists of frequent inspections, particularly where the frames are heated. Examine these every morning; water if necessary, remove faded leaves and wipe off the condensed moisture from the glass. Cuttings in cold frames should also be kept free from dead leaves, and should be watered sufficiently often to keep the compost just moist, but not wet. Shade should be given, particularly during the first few days, to prevent wilting. No ventilation should be given, except to lower a high temperature—over 27°C (80°F) is high for most hardy cuttings—until there are signs that rooting is beginning. From then onwards cooler and more open conditions should be allowed until the lights are removed altogether.

Of course many cuttings planted in cold frames during the autumn may not root until the following spring or summer. In such circumstances the frames should be kept closed during the winter. Ventilation is given and increased gradually in the spring as the weather gets warmer. Some cuttings, such as conifers, form a very hard callus, which appears to prevent root production. Paring this callus with a knife is usually beneficial in promoting rooting.

Semi-hardwood cuttings

Semi-hardwood (or half-ripe) cuttings are difficult to define and vary from those taken off shoots just beginning to mature at the base, to cuttings from shoots almost but not quite ripe. Experience has shown that the degree of maturity is important for some rather difficult plants, such as lilacs. It is impossible, however, to give the best date for taking cuttings of this type as it varies with the kind of weather in any season. For many shrubs the exact stage of maturity is not important, except that the more mature they are the longer they will take to root. The period for securing semi-hardwood cuttings is approximately from early July until the end of September. Shrubs successfully rooted at this period include lilacs, hydrangeas, spiraeas, cotoneasters, jasminums, pyracanthas, brooms and rose species. Evergreens including ceanothus, berberis, pernettyas, vacciniums, gaultherias, escallonias, evergreen azaleas and a number of conifers seem to root particularly well about this time. With most of these the cuttings consist of short lateral growths taken with a heel. Some types of semi-hardwood cutting root more readily if 'wounded'. This operation involves the cutting away of a thin sliver of bark from one side of the stem at the cutting's base. The stem can also be 'scored' with a sharp knife to produce the same effect.

The time taken for the various types of cuttings to root is, of course, affected by the conditions under which they are kept after insertion, and depends also on the species. On average softwood cuttings take from 2 to 6 weeks, the semi-mature type root in 5 to 25 weeks; hardwood cuttings rarely start rooting in less than 8 weeks and some take as long as 36 weeks.

Plastic tunnels Propagation in plastic tunnels is a method developed by the Glasshouse Crops Research Institute. Particulars of the construction of these tunnels were given in Chapter 2. This method is particularly suited to the amateur for the propagation of shrubs from semi-hardwood cuttings.

Before the tunnels are erected any necessary soil preparation should be done. This may include chemical sterilization to destroy weeds and weed seed. Dazomet is most convenient for the amateur and should be worked into the surface in accordance with instructions. Afterwards the soil surface is sealed by being covered with plastic sheeting and left for about two weeks. At this stage the seal is removed and two more weeks are required for aeration and to ensure freedom from toxic fumes.

On light soils a good dressing of peat should be mixed with the top few inches but on heavy land a liberal application of a mixture of sand and peat will improve the soil texture for cuttings.

This method is suitable only for semi-mature cuttings which can be taken from a wide range of shrubs from late June until the end of August. Deciduous subjects should be taken first to ensure rooting before the onset of winter, otherwise losses will occur. The average spacing of cuttings is 8 cm by 8 cm (3 in by 3 in). Most species root within about 4 weeks and thereafter ventilation should be given gradually by raising the sides of the tunnels.

Rooted cuttings are usually left in the tunnels over winter and are allowed to remain there without cover over the following growing season. At that stage they are usually well established nursery plants.

Root cuttings

Any part of a plant used as a cutting must be capable of producing new shoots. Normally shoots arise from buds found at the nodes or joints of stems, but buds may be formed accidentally, as it were, on other parts of the plant including stem internodes, roots and even on leaves. Such 'accidental' buds are described as being adventitious.

Adventitious buds appear naturally on the roots of such plants as raspberries and plums and give rise to suckers which, as explained in Chapter 8 are used in propagation. The roots of any such plants are also suitable for root cuttings. Moreover, there are other plants which, although they do not form adventitious buds naturally on their roots, can be induced to do so by ordinary methods of propagation. Seakale is a good example of this and most gardeners are only too familiar with the spectacle of pieces of weed roots, like dock and dandelion, sprouting and giving rise to more unwanted plants. The size of root cutting to take varies with the type of plant, but as a general rule relatively thick roots of a reasonable length are preferred. Thin, short pieces have very little food reserve and, if they grow at all, they take a long time to develop into vigorous plants. The propagating medium for root cuttings is generally similar to that used for other cuttings but, on the whole, root cuttings are less exacting in this and other respects than stem cuttings. This is because, as roots always grow in the soil, they are acclimatized to such conditions. Moreover, as root cuttings do not have leaves attached to them when inserted, humid conditions are unnecessary. It is found, however, that quite a number of species increased by root cuttings respond to mild bottom heat. On the other hand root cuttings of various hardy perennials develop readily into plants when inserted in ordinary soil in the open.

Seakale is a plant that can easily be propagated by root cuttings. Normally seakale plants are lifted for forcing when the leaves die down in the autumn. Surplus roots may then be trimmed off and used for propagation. Select pieces about the thickness of a pencil and cut them into 12 or 15 cm (5 or 6 in) lengths. The lower end should be cut slantwise but the cut across at the top should be straight. This acts as a guide in planting the right way up, which is very important (Fig. 43).

The cuttings (or thongs, as they are usually called), are tied in bundles, laid in damp soil or sand, and covered completely over until the spring. It will then be found that several buds have formed at the top end. All but the strongest of these should be rubbed off. The thongs are then planted with a dibber so that the top of each is about 2.5 cm (1 in) below the surface of the soil. The spacing is usually in rows 38 to 45 cm (15 to 18 in) apart with 30 cm (12 in) allowed between the thongs. Horseradish is another vegetable readily increased from root cuttings. The manner of preparation and planting is similar to seakale.

Among fruits, raspberries and their relations the blackberries and loganberries are easily increased from root cuttings. Suitable roots are 5 to 10 mm ($\frac{1}{4}$ to $\frac{1}{2}$ in) thick and are cut in pieces 8 to 10 cm (3 to 4 in) long. Such portions may be laid horizontally in a shallow trench and covered with a few inches of fine soil.

Various fruit tree root stocks of apples, pears, plums and cherries may be raised from root cuttings. Suitable roots may be secured by lifting

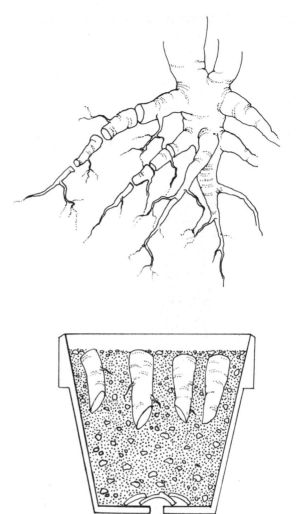

Fig. 43. Root cuttings. Quite a number of plants with thick, fleshy roots can be increased by root cuttings, the classic example being seakale. The upper drawing shows the roots of an established plant being severed. The lower drawing shows the cuttings inserted in a pot.

such trees in the autumn. Those between 5 and 15 mm ($\frac{1}{4}$ and $\frac{3}{4}$ in) are cut into pieces 10 to 15 cm (4 to 6 in) long. These are tied in bundles, laid in damp sand and are planted in the spring 5 to 8 cm (2 to 3 in) apart. Shoots arising from these cuttings should be reduced to one on each plant.

A number of trees and shrubs may be similarly propagated. These include several conifers, such as chamaecyparis, also poplars, the tree of heaven *Ailanthus altissima*; the sumach *Rhus typhina*, and *Daphne genkwa*.

Various herbaceous perennials are readily increased by root cuttings. A number of these including *Anchusa azurea* and its varieties, perennial verbascums, eryngiums and oriental poppies, may be planted in the open in the same manner as seakale. The ordinary garden phlox, *Phlox paniculata*, is often increased by root cuttings and this method is quicker than dividing the roots. It consists of selecting the stronger roots and cutting them into pieces about 5 cm (2 in) long. These are sprinkled on a sandy compost contained in boxes or pans and covered with about 1 cm (½ in) of the same medium. Place the receptacles in a frame or greenhouse and, when the shoots are well through, the young plants should be pricked off in a cold frame or in a nursery bed in the open.

Several alpines may be propagated from root cuttings; the little tufted plant *Morisia monantha* provides an example. Plants are lifted in June after flowering and the thicker roots cut into lengths of about 2.5 cm (1 in). The fibrous roots are trimmed off. Insert the cuttings 2.5 cm (1 in) apart in pans with their tops just level with the soil surface. Place the receptacles in a frame, or if in the open cover with a sheet of glass. In about a month a rosette of leaves forms at the top of each cutting, and they may then be potted in small pots. Other alpines that can be propagated from root cuttings include *Pulsatilla vulgaris*, *Primula denticulata*, alpine geraniums, erodium, phlox, campanula and mertensia.

It has previously been asserted that with vegetative propagation the new plants are always exactly like their 'parents'. It is known, however, that root cuttings do not always give rise to plants that are true to type. Thus when variegated pelargoniums are propagated from root cuttings they usually lose the variegation.

Leaf cuttings

Adventitious buds arise on the leaves of certain plants, making leaf propagation a possibility in such cases. The natural production of buds is very rare but bryophyllum is an example where this occurs, giving rise to new plants. The leaves of certain other plants may be induced to form adventitious buds and roots.

Broadly speaking, propagation from leaves

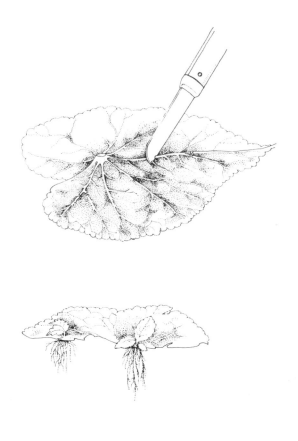

Fig. 44. Leaf cuttings. *Begonia rex* can be propagated by pegging its leaves on to sandy compost. It is usually necessary to cut the mid-rib and the main veins of the leaf to induce rooting. A new plant may be produced at each slit.

takes two forms. The first is exemplified by the well-known *Begonia rex* which has large ornamental leaves. In this case the leaf veins are cut through at several points (Fig. 44). The leaf is then spread out and pegged down on damp, light compost. If enclosed in a warm frame roots and buds are produced at the incisions, each giving rise to a new plant.

A second type of leaf propagation consists of inserting the leaf stalk into the compost like an ordinary cutting. Two well-known rock plants, haberlea and ramonda, provide examples of plants that can be increased by this method. Strong young leaves are preferred and it is essential for a bud to be present in the leaf axil; otherwise they cannot grow. The leaves are inserted to a depth of about 4 cm (1½ in) in peaty compost placed in a well-shaded frame. Leaves of other plants treated similarly for propagation are streptocarpus, *Saintpaulia ionantha* (Fig. 45) and lachenalia. *Begonia rex* may also be treated in this way, but the leaves are cut into strips tapering towards the base, each strip being inserted vertically. Another plant which grows readily from leaf pieces is sansevieria.

Shrubs which may be increased by leaf cuttings are *Viburnum rhytidophyllum* and *Eriobotrya japonica*. It is interesting to note that the leaves of the climbing shrub, *Hoya carnosa*, produce roots readily but fail to develop buds, so that no further advance is possible.

A heated propagating frame with a lid is normally necessary to root leaf cuttings, but the strong-growing species of sedums provide a marked exception to this rule, for their leaves will root easily in the open. Indeed, this applies to

Fig. 45. Leaf cuttings of African violets need to be treated rather differently from those of *Begonia rex*. Instead of cutting the mid-rib, the stalk of the leaf is inserted into the sandy compost, and tiny plants will appear at the point at which the leaf stalk goes into the soil.

almost any part of the plants. Thus, if we take a sedum plant, any time between April and September, break it up and then sow the leaves and stem pieces over the soil, many plants will be produced.

Leaf-bud cuttings

As distinct from leaf cuttings, leaf-bud cuttings (similar to those used for budding) may with certain species be induced to root. The most important plants sometimes increased in this way are loganberries and blackberries. From July to September well developed shoots of the current year are selected to provide buds. The buds are removed by inserting the knife about 1 cm (½ in) below the bud and bringing it out about the same distance above the bud. A fairly thin slice should be removed without cutting into the pith of the stem.

The leaf buds are planted in a sandy compost in a cold frame, each being inserted just below the surface with the leaf intact and left exposed. The lights are then placed in position and, if open to the sun, should be shaded. Water to maintain reasonably moist conditions.

In from 6 to 8 weeks the cuttings should be rooted. Ventilation is then given and gradually increased until finally the lights are removed altogether. As a rule the cuttings are left until the spring, when they may be planted out in nursery beds.

A similar type of leaf-bud cutting is sometimes used in the propagation of camellias (Fig. 46) and *Ficus* (Plate 3). In this case the buds are inserted around the side of a 9 cm (3½ in) pot. They are held in position just above soil level with a piece of wire bent in the shape of a hairpin. Another method which is now preferred for increasing camellias is to cut off a piece of mature stem about 2.5 cm (1 in) long, which is provided with a leaf and bud near its top. About 5 mm (¼ in) from the base of this cutting an upward slanting cut is made extending one-third to half-way through the stem. This nick appears to promote the production of roots near the base. Insertion should be done firmly around the side of a 9 cm (3½ in) pot.

Transplanting and potting

Alpines are usually potted when rooted and are kept in a shaded frame for a few days. In the pots they are allowed to establish themselves before

Fig. 46. Leaf-bud cuttings. Certain plants such as camellias can be propagated by leaf-bud cuttings. Take a cutting of half-ripened stem with leaf attached, and with a growth bud in the axil, as shown. Plant the cuttings around the edge of a pot as indicated in Fig. 39.

being planted out. Various rooted shrubs may be planted outside in nursery beds in a shaded sheltered position. Mulching with leafmould, peat or spent hops is often beneficial, and water should be given if necessary. Some shrubs, such as brooms and gorse, are difficult to transplant. These should be potted from the cutting bed and then may be plunged in peat, sand or gravel out of doors. This method allows them to develop into strong plants before being set in their permanent positions.

Stock plants

In nurseries where large quantities of cuttings, layers, scions and buds are required, it is usual to have a number of stock plants to supply these. Hardy species, such as blackcurrants and gooseberries, are usually planted in groups to form what are called stool beds. These beds produce a crop of stems every year for propagation. Stock plants of various species, including hardy kinds, are often grown in pots so that they may be transferred into a heated greenhouse and forced into rapid growth to provide propagating material when required. Examples of plants treated in this manner are roses, hydrangeas and fuchsias.

12
The Techniques of Grafting and Budding

Grafting is the art of inducing a piece of the stem of one plant, called the scion, to unite with the rooted portion of another, known as the stock. The two so united grow together to form one plant, yet each maintains its individuality and a shoot arising from one is distinct from that produced by the other. Normally, however, the role of the stock is to provide the plant's roots while the scion furnishes the top growths.

The primary object of grafting and budding is to reproduce plants that are more difficult or cannot be propagated at all by other vegetative methods. Thus, the various tree fruits, such as apples, pears and plums, are extremely difficult to root from any type of cutting, while by grafting large numbers can be raised with comparative ease.

Another important reason for grafting is to enable certain plants to be grown on roots other than their own, and this is often advantageous. Tree fruits again provide an example and, in the case of apples, several distinct root stocks are used with the object of securing trees suitable for different purposes and conditions. One stock, for instance, called Malling 9, produces a dwarf tree that comes into bearing early in its life; Malling 7 stock gives a medium-sized tree, while trees grafted on to Malling 16 grow vigorously. (See also pages 149 and 150.)

Roses are usually budded (a form of grafting) although most varieties root easily from hardwood cuttings. However, when some varieties are grown on their own roots they make poor weak plants but, when budded on stocks of *Rosa canina* or *R. laxa*, strong vigorous bushes are produced.

Clianthus formosus, the glory pea, is not easy to cultivate and rarely remains alive for long under cultivation if grown on its own roots. But when grafted on seedlings of *Colutea arborescens* (bladder senna) the borrowed roots confer upon it a new lease of life and it grows vigorously and flowers freely.

It will be seen that although a scion never loses its identity whatever the stock it is grafted upon, yet the latter has considerable specific effects on the scion. The most important of these is that relating to vigour already mentioned, but it is interesting to note that a stock which dwarfs a particular species may have the opposite effect on another species. Thus the quince stock *Cydonia oblonga* when used for pears, tends to reduce their size, while hawthorn *Crataegus monogyna* grows larger when grafted upon quince.

The statement that a grafted scion never loses its identity is subject to qualification. Actually in very few instances peculiar composite plants known as 'graft-hybrids' have arisen through grafting. Such a plant has an internal core of tissue derived from the scion while the outer layer originates from the stock. Perhaps the best known graft-hybrid is + *Laburnocytisus adamii*, which has pinkish-purple flowers. It is a graft-hybrid of the common laburnum and the purple broom *Cytisus purpureus*.

The stock also affects precocity, that is, the age at which the grafted plants come into production. With fruit, quality and yield are influenced. Sometimes the stock exercises some control over the shape of the tree and its hardiness.

Another important use of grafting is the speedy

substitution of one variety of plant for another. For instance, if one has a mature tree of a particular variety of apple, say 'Bramley's Seedling' and wishes to change it to 'Cox's Orange Pippin', this transformation can be effected by what is called framework grafting and the reconstituted tree will be in full production in a few years.

Grafting has several disadvantages. Thus, grafted plants may be prone to suckering, that is, to producing unwanted shoots from the rootstock. This is a bad fault with some plum stocks, and can be a serious nuisance. Sometimes the union between the stock and the scion is imperfect, resulting in breakages or poor nutrition. As a general rule it is preferable to have a plant on its own roots unless growth is known to be unsatisfactory in such circumstances, or the plant is difficult to increase by means other than grafting. It is probable that nurserymen use grafting and budding to a greater degree than is necessary. This is because these methods lend themselves to the mass production of plants.

The prospects of success with grafting are, however, enhanced if great care is taken to use suitable stocks for each species. Stocks very prone to suckering, such as privet, which is sometimes used for lilac, should be avoided. Apart from this, as a general rule the closer the relationship between plants, botanically speaking, the more successful is the union between them likely to be. It is only common sense to assume that a holly grafted on an apple or vice versa is not likely to thrive.

Botanical relationships between two species, however, is not a sure guarantee that they will unite successfully, and in the long run experience provides the most reliable guide. When two given species cannot be successfuly grafted together they are said to be incompatible. Incompatibility is a matter of degree and varies from slight ill-health of the scion to a total failure of the union to knit. As a general rule particular varieties of a species are grafted on to seedlings of that species or a related species. Thus, named varieties of lilacs are grafted on seedlings of the common lilac, special forms of laburnum on to seedlings of the common laburnum and the coloured hawthorns on to seedling quickthorns. In the case of fruit trees, however, seedlings are rarely used today as rootstocks. This is because trees grafted on to seedlings are not always uniform in size and other characters, so that certain standard stocks

which are vegetatively propagated are preferred.

Whatever the method of securing stocks for grafting, the propagator must ensure that he has adequate supplies of these suitable for his purpose. As a rule stocks for grafting in the open are planted in the soil and for most subjects such as fruit trees, or roses, should be established at least a year before grafting or budding is attempted.

Grafting out of doors is usually undertaken in early spring just when growth is beginning. The scions, however, should be severed from the parent plant when they are quite dormant. Usually they are then bunched together and laid with their bases in damp sand or soil in a shady position.

Success with the operation of grafting depends entirely on how well the cambium of the stock and the cambium of the scion are brought together, for no union can occur unless this is achieved. Even when only a small area of the respective cambiums meet, the operation may be successful, but quicker and often more satisfactory results are ensured when the maximum of firm cambial contact is secured. This, therefore, should be the aim, and it is facilitated by making firm, clean cuts with a really sharp knife in such a way that the parts fit snugly and firmly together.

Scions vary in length according to the type of plant, but a common length is about 15 cm (6 in). The stocks should be well established and sufficiently strong to sustain the grafted plant. The more important forms of grafting are described below.

Splice or whip grafting This is one of the simplest methods of grafting (Fig. 47). It is most suitable when the stock and scion are approximately the same thickness. First of all the stock must be cut back to the point where it is intended to graft. A long slanting cut is made at the base of the scion, and one of a similar slope to the stock apex. The cut edges of the two stems are then simply placed together and tied firmly in position with plastic tape or raffia. If the stock is slightly thicker than the scion, the latter should be kept flush with one side of the former so that cambial contact is ensured on one side at any rate.

Whip and tongue grafting This is simply an improved form of splice grafting and is probably the most widely used method, and the normal form of grafting in the propagation of young fruit trees. The stock is usually cut down to within a few inches of ground level and both stock and

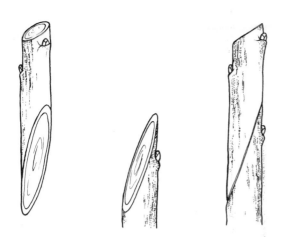

Fig. 47. The splice or whip graft may be used where stock and scion are of equal thickness. The cut surfaces should (as with all grafts) fit together well.

scion are prepared as for splice grafting. An additional cut is made, however, downwards about the centre of the slanting cut at the stock apex to form an upward-pointing tongue (Fig. 48). A corresponding tongue is made on the sloping cut at the base of the scion. When the stock and scion are brought into contact, the tongues overlap and not only hold the separate parts firmly together but also ensure good cambial contact. If the stock is appreciably thicker than the scion it is best to pare off just a portion of the rind (bark) so that the cambium on both stock and scion meets.

Rind grafting This form is suitable for grafting a stock which is very much thicker than the scion (Fig. 49). It is a simple method and can be thoroughly recommended to the amateur. The stock is cut across transversely with a saw and a clean slit about 5 cm (2 in) long is made in the rind downwards from the apex. A similar length is prepared at the scion base by paring down one side, and this portion is then pushed underneath the stock rind where the slit has been made. The

Fig. 48. Whip and tongue grafting. *Top*: cutting the stock, and the prepared stock. *Centre*: cutting the scion – upwards, from just under a bud. *Bottom*: the two should fit neatly together, and should then be bound together with adhesive tape. When raffia is used, the union should be sealed with grafting wax.

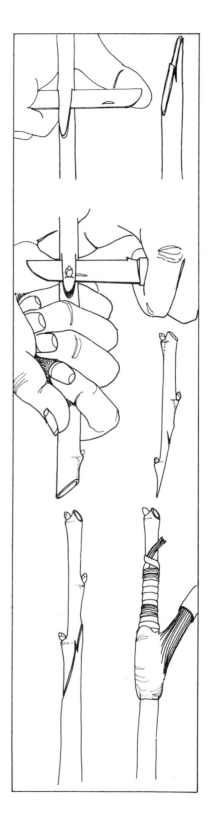

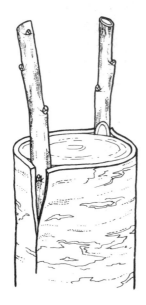

scion is then secured by tying with plastic tape. This method is often used for top-grafting established fruit trees.

Saddle grafting The stock and scion should be approximately equal in diameter. The former is prepared by making a short cut on either side at the apex to form the shape of an inverted 'V' (Fig. 50). A corresponding incision is made at the base of the scion so that it fits saddle-fashion on the stock. This type of grafting is often used for rhododendrons, the operation being performed in spring under glass, using softwood scions.

Wedge graft This involves preparing the base of the scion in the shape of a short wedge which fits into a similarly shaped cleft in the stock (Fig. 51). Both parts should be about the same thickness.

Fig. 49. Rind grafting is useful where the stock is much thicker than the scion. The scions are inserted under the bark and bound into place.

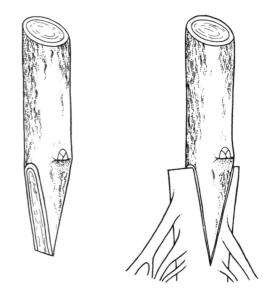

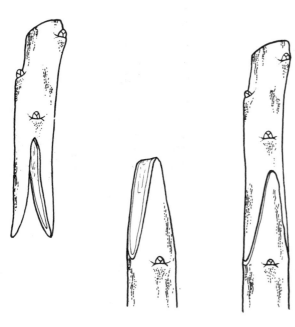

Fig. 51. Wedge grafting is usually carried out very low on the stock – almost at the point where roots arise.

The forms of grafting described above involve cutting off the top of the stock, but with the methods mentioned below this is not necessary.

Inlaying This simply involves making an incision in the stock and replacing it with a portion of the scion cut to precisely the same shape. The scion is prepared by making two equal cuts at its

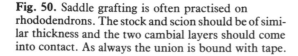

Fig. 50. Saddle grafting is often practised on rhododendrons. The stock and scion should be of similar thickness and the two cambial layers should come into contact. As always the union is bound with tape.

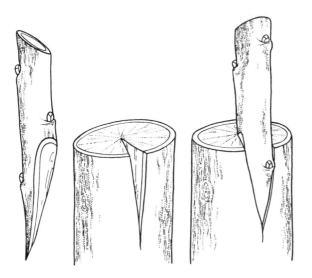

Fig. 52. Inlay grafting is practised where the stock is larger than the scion. It may be used on herbaceous plants, and if the scion is a good fit, tying will be unnecessary.

Fig. 53. Veneer grafting involves the removal of bark from the thick rootstock and the insertion of a matching portion of scion wood, the back of which has been shaved away. Tack the scion into position with panel pins.

base to form a long sideways wedge (Fig. 52). A similarly shaped piece of rind and wood is then removed from the stock, into which the scion is inserted. Stocks used for inlaying should be thicker than the scions.

Veneer grafting This is a similar type of grafting to inlaying, except that the rind only is removed from the stock (Fig. 53). The latter should be much thicker than the scion. A modification of veneer grafting is to remove a narrow strip of rind from the stock. A fairly short scion is cut off above a bud, and after shaving well down on one side, the whole scion is inserted into the opening in the stock.

Side grafting This method consists of the insertion of the scion base for an inch or so under the rind on one side of the stock (Fig. 54a–b). There are several modifications of this method such as preparing the stock as for shield budding and then inserting a wedge-shaped scion for about an inch beneath the bark. Sometimes the stock is prepared to receive the scion by driving a thin chisel, screwdriver, or knife blade into the bark. The end of the scion is cut into a thin wedge and forced into the incision until the cut surfaces are covered by the bark of the stock. Side grafting is a popular method and is often used for shrub grafting, particularly for dwarf conifers such as certain species of pinus and picea.

Framework grafting Varieties of apples and other kinds of fruit if unsuitable may be substituted on established trees with superior varieties by grafting. Two methods may be used, namely 'top grafting' or 'framework grafting'. The former involves cutting the trees down to a low level and inserting apical grafts on the main branches. A tree treated in the manner has therefore to rebuild its framework before again coming into full production, and this may take many years. With framework grafting (Fig. 55a–b), however, the main framework is left so that, in a few years after grafting the new variety, the tree is again in full production.

In preparing for framework grafting the tree is first pruned into shape. Side branches are thinned out, but as many as possible of the laterals which are less than 2.5 cm (1 in) in diameter are left. Scions are grafted on these small laterals by the method called 'stub grafting' which obviates the need for tying. The scions having six or seven buds are prepared by making two sloping cuts at their bases. One cut is about 2.5 cm (1 in) long

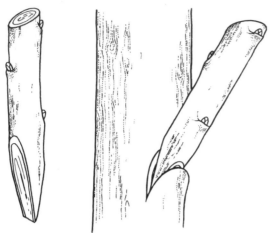

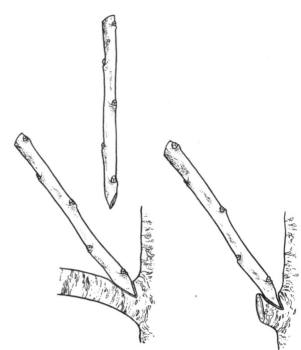

Fig. 54a. Side cleft grafting is useful where stock is thicker than scion. The scion may be more readily inserted if the stock is bent over to open the cut, and allowed to spring back afterwards.

Fig. 55a. Framework grafting involves the replacement of fruiting branches on trees where the variety is to be changed. Stub grafting is one of the methods used. After bending the young branch down so that the cut is opened to take the scion, it is released and cut off.

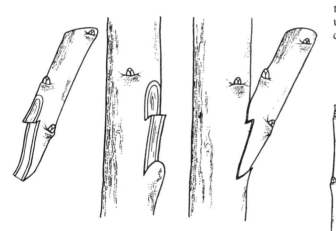

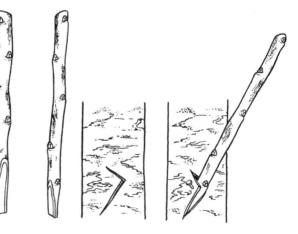

Fig. 54b. Side tongue grafting is used where the stock is only slightly thicker than the scion.

Fig. 55b. A shortage of branches for stub grafting may make it necessary to insert some inverted 'L' grafts under the rind on larger branches. Such grafts may be held in place with small nails as shown.

and the other opposite to it and about half that length, to form an uneven wedge. The laterals are prepared to receive the scions by making a slit beginning about 1 cm ($\frac{1}{2}$ in) from their base and sloping downwards towards the main branch penetrating to about the lateral's centre. On pressing the lateral away from the main branch, the cut opens forming a gap for the prepared scion, which is held naturally in position by the spring-like tension of the lateral when it is released. The lateral is then cut back just above the scion and the union bound with tape. Whip and tongue grafting may also be practised on the laterals.

Should there be insufficient laterals to permit a branch being properly furnished with grafts, perhaps 30 cm (1 ft) or less apart, scions may be inserted under the bark by various methods of side grafting. One such method is the 'inverted L graft'. This consists in making a cut through the bark in the shape of an inverted 'L'. The lower part of the cut is 2.5 to 3 cm (1 to 1$\frac{1}{2}$ in) long, with the upper one somewhat shorter and made at an obtuse angle to the first. The scion is prepared by making a sloping cut at its base on one side and a shorter cut made opposite but somewhat to one side. This allows cambial contact on both sides of the scion when it is inserted. It is held in position with a thin flat-headed nail driven through the raised rind and the scion base into the wood of the branch. The union is then bound with tape.

By these methods 150 to 200 scions may be grafted on a medium-sized tree. During the summer a grafted tree must be gone over several times and all new growths, other than those from the scions, removed. In grafting fruit trees, care must be taken to use varieties that will ensure effective pollination. Frameworking is successful with apples, pears, plums and cherries and also with several nut fruits.

Tying and sealing grafts With most methods of grafting it is necessary to tie the scion firmly to the stock and polythene tape (Fig. 56) is often used for this purpose, though raffia was very popular at one time. Sometimes, as with the inverted 'L' graft, the grafts are held in position by a small nail. After tying, any unions which are open at the top are usually sealed around with some adhesive substance to prevent the entry of water and to check drying out of the wounded tissues.

Grafting wax is sometimes used and there are several different recipes for its production. The

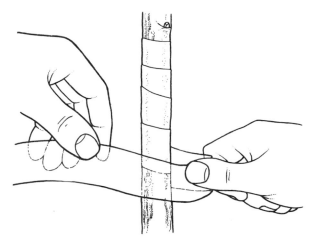

Fig. 56. Diagram showing how the scion is bound firmly to the rootstock with transparent polythene tape.

following is an example: 1 part tallow, 2 parts beeswax, 4 parts resin. These ingredients are melted and well mixed together. The whole is then poured into cold water, and kneaded with the hands until it becomes white and assumes the consistency of soft putty. This mixture, in a warm but not hot condition, should be applied to the union with a paint-brush. Certain waxes may be purchased ready for use, and some may be used cold. Petroleum jelly is also used for sealing grafts and is fairly effective. In recent years adhesive tape is being used both as a tie and to act as a substitute for wax.

Bench grafting
This method simply consists of grafting scions into bare root stocks at a bench. The work is much less laborious than grafting stocks planted in the soil as one is free from the hazards of weather.

Bench grafting is a popular method for various named varieties of ornamental trees and shrubs such as laburnum, flowering apples (malus), flowering plums (prunus), sorbus, syringa and amelanchier. Fruit trees may be grafted in this manner and it is also used in the grafting of roses for the production of flowers in commercial greenhouses.

One or more of the grafting methods already described may be used for bench work but the simple whip graft (without the tongue) or the side graft are the most popular. Tying and waxing are as done in the open. The grafted plants may then be stood in moist peat for planting out when convenient.

Grafting under glass
In nurseries a considerable amount of grafting is done in greenhouses. This allows better control over the environment, and there is less risk of the graft drying out before the union is effected. Under these conditions the application of wax to the union is usually unnecessary. Another advantage of indoor grafting is that the work can be done at times when it would be impracticable out of doors. Winter is the usual period for grafting under glass, but some plants are grafted in late summer.

Stocks for use in grafting shrubs are usually grown in pots, and root activity should be well advanced at the time of grafting. The most popular methods used under glass are splice grafting, wedge grafting, side grafting and whip and tongue grafting. Rhododendrons are usually saddle-grafted. Whatever the method, the grafts should be secured firmly and then placed in a propagating frame until the union is effected. Afterwards freer ventilation is allowed until the plants may be stood in frames in the open.

In greenhouses the scions of certain plant varieties such as clematis, wisterias and begonias are grafted on to roots of seedlings of these species. Suitable roots are secured when the plants are dormant. Simple splice grafting is generally used. Clematis varieties are grafted on the roots of the wild clematis or old man's beard *C. vitalba*. The method is to make scions from soft young stems of plants specially forced into growth for this purpose. First, a piece of stem about 2.5 cm (1 in) long and having a node near its apex is cut off. This is then slit in two down the centre to form two scions each having a bud. The stock is prepared by paring off one side of it near the apex. Here the scion is fitted and tied in position. The grafted plants are potted into small pots with the union slightly below soil level, and the pots are stood in a propagating frame. In this case the root stock acts as a nurse and keeps the scion alive until it roots into the soil. Nowadays most clematis are propagated by cuttings rather than by grafting.

The majority of herbaceous plants may be grafted successfully, but, as most of these can be increased by other, simpler methods, grafting is necessary only with a limited number of plants. An example is the double form of gypsophila, which is grafted on pieces of root of the ordinary single type, *G. paniculata*. The usual method is a form of inlaying. Young shoots of the double gypsophila are secured from plants grown in heat. The base of each of these is shaped as a kind of side wedge and is inserted into a similarly shaped incision made in the piece of rootstock which is an inch or so long. The grafted roots are usually dibbled into sandy compost and kept in a propagating frame. Recently, however, certain nurserymen have shown that gypsophila varieties can be easily increased from cuttings.

Herbaceous grafting is also used for the propagation of certain trees and shrubs which are difficult by other methods. In this case soft unripened scions are used, and the grafted shoots of the stock should also be immature. Certain varieties of pines and walnuts are grafted in this way, wedge grafting being preferred for this purpose.

Inarching
A distinct form of grafting is called inarching or grafting by approach. It is often done indoors, but can also be performed in the open at any time when growth is active. This aims to secure a union between the stock and the scion before the latter is severed from its parent. Inarching therefore bears a similar relationship to grafting as does layering to propagation by stem cuttings.

In order to perform this operation it is essential to have at least one of the two plants in a pot so that the stems to be united can be brought into contact. The simplest form of inarching is to shave off the bark from each stem and then fit and tie the wounded parts firmly together. When the union is effected the scion is severed from its roots and treated as an ordinary graft. Vines and nuts are examples of plants which may be propagated by inarching. One of the most commonly used forms of inarching is called cleft inarching (Fig. 57).

When grafting is successful the buds soon start to grow. If several on each graft develop into shoots it is necessary in the case of trees to reduce them to one only. Disbudding should not be done, however, until the shoots are far enough advanced to select the best and most suitably

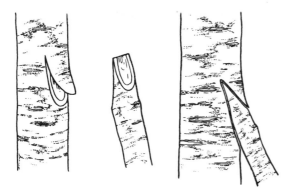

Fig. 57. Cleft inarching unites stock with scion before the latter is severed from its roots. One of the plants must be pot grown if this type of graft is to be made possible. It is usually carried out when the plants are in active growth.

placed one to be left. Any buds that grow on the stock should also be rubbed off. It is best to do this gradually and it may be an advantage to allow shoots from the stock to develop until the scion shoots are sufficiently advanced to maintain a balance between root and top growth.

Budding, a form of grafting

Although budding is usually regarded as a distinct method of propagation, the principle is exactly the same as in grafting. With the latter method a complete length of stem is used as the scion, whereas budding consists of grafting a piece of rind, with bud attached, on to another stem or root. The union takes place between the cambium layer attached to the piece of rind and the cambium layer of the stock. The bud is there to provide the first shoot of the plant but takes no part in the actual union. Some of the disadvantages often attributed to grafting, such as suckering and the possibility of using unsuitable stocks, apply of course to budding.

Budding is very popular with nurserymen for the following reasons: Firstly it is a quick and efficient method of propagation. Thus in the case of roses a skilled budder can bud as many as 100 plants per hour with the probability of over 90% of these being successful. Secondly budding is often the best of all methods to secure the maximum number of plants from a given quantity of propagating material. Practically every bud on the parent plant may give rise to a new individual. Budding is therefore of particular value in rapidly increasing new and often valuable varieties of plants, such as a new rose seedling of promise.

This operation is confined mainly to raising woody plants but can be carried out successfully on some herbaceous perennials. Roses are the most important group of plants that are increased mainly by budding (Fig. 58), but it is also widely used in the propagation of fruit trees. In nurseries it is a common practice to bud established fruit tree stocks in the summer and those that fail to take are grafted the following spring. Budding, however, is usually preferred to grafting in the case of stone fruits. This is because such fruits are liable to a complaint called gumming when a large wound is made in the stem. Should this occur, sap oozes from the stem in the form of a gummy substance. This may continue for a long period, and healing of the wound is retarded. Many ornamental trees and shrubs may also be increased by budding, including flowering apples, flowering cherries, named varieties of hawthorns, acers, elms and horse-chestnuts.

Budding is generally done at the height of the growing season, that is from June to August. This is because during these months the rind of most plants separates easily from the wood and facilitates the operation. Stock plants established on the site for a least a year are used for this purpose. Before starting to bud, any laterals near the stock base should be trimmed off to leave a clean stem for the bud.

As a rule the bud is inserted near the ground, and in the case of roses almost on the root. This tends to obviate the production of suckers and encourages the scion to produce its own roots, which is desirable with roses and other ornamental plants. With fruit trees, however, scion rooting is most undesirable because the usual beneficial effect of the stock on the tree may be lost or reduced if the scion produces its own roots. When buds are inserted near the soil this method is described as bottom working. Top working means budding at the top of a tall stock which provides the main stem, as in standard roses. In this case the bud may be inserted on the stem but, if the rind of this is very thick, it is preferable to bud at the base of young lateral shoots. As a rule several buds are used to provide shoots for the head of standard roses and suchlike plants.

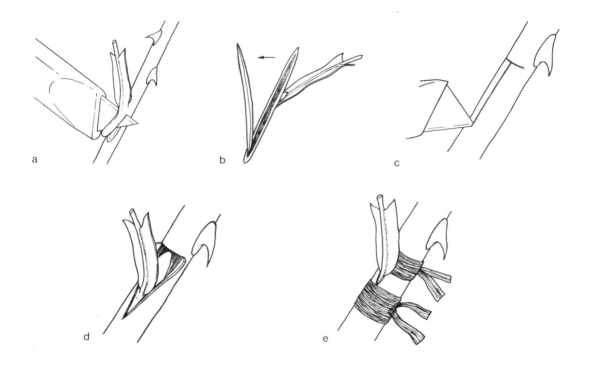

Fig. 58. Budding roses (shield budding). (*a*) Take a well developed bud from halfway up a strong stem which has flowered. (*b*) Remove the thin sliver of heartwood from the back of the bud. (*c*) Make a T-shaped incision in the stock. (*d*) Lift the corners of the bark and insert the bud. (*e*) Cut away any protuding shield rind and bind with tape or with a rubber budding patch.

Young stems of the current year's growth provide the buds, and if there is a fair amount of budding to do a supply of these should be secured before starting. After being severed the stems should not be left in strong sun and other measures such as standing them in water may be necessary to prevent drying out. For this reason also dull, moist weather is preferable for budding, and the buds should be inserted in the stock immediately they are isolated. Select plump wood buds from medium-sized stems and do not use those of the immature soft tips.

There are several methods of budding but by far the most important is called 'shield' or 'T budding'. This is a simple operation which with a little practice anyone should be able to perform successfully. A good sharp budding knife is essential. The stock is prepared by making a clean vertical cut about 2.5 cm (1 in) long right down to the wood, but no deeper. At the top of this make a transverse cut to form a 'T'-shaped

incision. A bud is then sliced from the stem with about 1 cm ($\frac{1}{2}$ in) rind on either side of it.

There is always a great deal of controversy as to whether or not the wood adhering to the rind should be pulled out. As a rule it is better to do so because this allows improved cambial contact. In removing the wood, however, there is the risk of pulling out the base of the bud, which is a little green structure about the size of a pinhead. If this is removed a small hole will be seen directly underneath the bud, which should then be discarded. Sometimes, however, a tiny hole can be seen just below the minute protuberance at the bud base. This really indicates the end of the leaf-stalk and does not affect the condition of the bud.

If the risk of spoiling the bud is considered too great, the wood attached to the rind may be left alone, but in that case it is particularly important to cut off the bud with a fairly thin slice. The attached leaf on a bud should be cut off so as to reduce transpiration to the minimum, but the leaf-stalk or petiole is allowed to remain and serve as a handle.

When the bud is ready the rind is carefully and cleanly lifted from the wood on either side of the incision on the stock. Into this the little shield held by its tiny handle is inserted from the top. Extra care should be taken to ensure a neat job and the likelihood of success. Speed comes with experience. To complete the operation, fasten a rubber budding patch around the stem, or tie the bud in place with polythene tape. This covering prevents drying out and the entry of rain or insects. Some authorities assert that the buds should preferably be inserted on the north side of the stocks, but experiments have shown that there is no advantage in doing so.

Another type of budding is called 'patch budding' (Fig. 59). This involves the neat cutting out of a patch of rind, usually in the shape of a square, from the stock. The patch removed is then replaced by one of the same shape and size from the scion, which must have a bud near its centre. For patch budding the rind of the stock and of the scion should be of similar thickness. The substituted rind is neatly taped in position.

Buds inserted in summer may produce shoots the same season, but the majority remain dormant until the following spring. A few weeks after budding it is usually possible to say whether or not the bud has taken. If it looks fresh and plump all is probably well. Another sign of success is the natural fall of the leaf-stalk, which usually withers but which remains attached when the result is negative. To allow the bud to swell the tape should be cut as soon as it is evident that they have taken. Rubber budding patches will expand and perish as they age so there is no danger of constriction.

Normally, buds are inserted in stocks which have the top growth intact. In the case of roses the stocks are cut back as closely to the union as possible in the spring. With fruit, however, and some flowering shrubs, it is usual to leave a piece of stem a few inches above the bud, called a snag. This is done to form a temporary stake for the young shoot to which it is tied. The snag is usually cut off about a year later, as closely as possible to the new shoot to allow it to grow straight from its base. As with grafting, any growths from the stock should be kept trimmed off.

Chip budding
The usual method of budding known as 'shield' or 'T' budding is a rather skilled and delicate operation, with sometimes disappointing results. Chip budding is an alternative method which may be tried if previous results have failed.

Fig. 59. Patch budding involves the removal of a square of rind from the plant that is to provide the shoot material. It should fit exactly into a gap left by the removal of a piece of rind on the stock. It may be used to insert a bud from a male plant on to a female, so ensuring the production of berries as a result of ferilization (e.g. practised with *Hippophae rhamnoides*, the sea buckthorn).

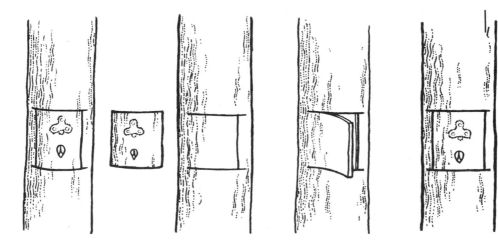

Chip budding (Fig. 60) simply involves the removal of a chip of stem from the stock plant and its replacement with a similar piece from the scion. The new piece must, of course, have a bud attached. Chip budding is effected by making two cuts across the stock about 3 cm (1½ in) apart. Next remove the chip of stem between the cuts with a knife and replace it with a similar chip from the scion. The bud should be near the centre of the new chip. Finally, bind the bud chip to the stock with overlapping polythene tape.

Fig. 60. Chip budding is easier to perform than shield budding. It can be performed when the bark will not part readily from the wood – usually in early spring before growth commences.

Plate 1. (*a*) Individual peat blocks are especially useful for large seeds such as sweet corn.
(*b*) This heated propagator base is large enough to accommodate three individual seed trays.
(*c*) This heated propagator has a moulded clear plastic top with adjustable air vents, specially designed to ensure that condensation is directed to the edges of the tray. The propagator is thermostatically controlled and provides even heat distribution between 7°C (45°F) and 30°C (85°F).

a

b

c

a

b

c

d

Plate 2. Polythene wrap propagation allows a large
number of semi-ripe cuttings to be propagated in a small
space. After being prepared and dipped in hormone
rooting powder the cuttings are laid on a strip of moist
sphagnum moss and the polythene folded over their bases
and rolled up. The cuttings are best placed in a
propagating frame.

13
Layering Plants

Layering is a method of plant propagation which is designed to induce plant stems to produce roots while they are still attached to and sustained by the 'parent' plant. It is a reliable means of increase and is often adopted for species which are difficult or impossible to root from cuttings. Layering does not lend itself to the production of large numbers of plants and is, therefore, used by the nurseryman only for the more difficult species.

This method is of particular advantage to the amateur who does not require a large quantity of any particular species, for layering is comparatively easy to perform and does not require the facilities necessary for certain other methods of propagation, such as softwood cuttings. Layering is used largely for increasing woody plants, but a few of the herbaceous type, notably border carnations, can be propagated in this way.

Ordinary or simple layering

The method usually adopted for dealing with trees and shrubs is referred to as ordinary layering. In nurseries it is usual to establish a special 'stool bed' for the purpose which should preferably be sited in a sunny sheltered position on well-drained light soil. The permanent plants are well spaced out to provide ample room for layering.

Stock plants are not allowed to form a trunk or main stem but are cut down to not more than 30 cm (12 in) from ground level. The object is to induce the production of young shoots from the base. These are most suitable for layering and, being near the soil, are easy to handle. The establishment of a good stool bed takes time, but once it has been achieved the reward is an annual crop of strong young shoots for layering which may remain in production for many years.

The soil for layering around the stock plants should have liberal applications of peat and sand to make it porous and yet retentive of moisture. Weeds should be kept down by cultivation as necessary. Autumn is the most suitable time for this type of layering, but the work often continues in open weather throughout the winter.

Ordinary layering (Fig. 61) may consist of no more than bending the shoots downwards and covering the portion of the stem nearest the ground with soil. Very often, however, the shoots to be layered are cut about half-way through the stem and are slit for a short distance in an upward direction. Sometimes a notch is made in the part to be buried or a complete ring is cut around the shoot. The stem is then bent at the point of wounding to form a right angle, and in this shape is pressed down and covered firmly with soil leaving the tip standing upright.

Layering on a large scale should be done systematically. First, the shortest stems near the base of the stock plant are dealt with. Continue to work outwards, spacing the layers as evenly as possible and making each quite firm.

The amateur gardener can easily layer a few shoots of any shrub in his garden, provided the stems can be brought down to the ground. It is essential to use young shoots only. Older branches are usually slow to root and rarely develop into well-shaped specimens. A notch, slit, ring or even the paring off of some of the bark

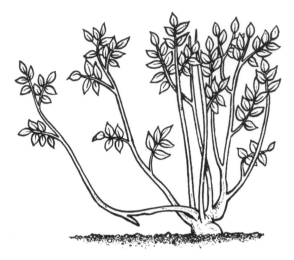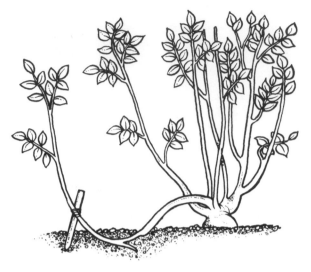

Fig. 61. Layering. In the left-hand drawing the lower left branch has been prepared for layering by slitting. As the branch is bent downwards into the soil the slit is opened and the sharp curvature on the uncut part of the stem impedes the flow of sap, thereby encouraging rooting. The right-hand drawing shows the layer firmly placed into soil and firmly staked, to prevent movement by the wind.

on the part to be layered is usually beneficial. Fix the cut portion firmly in the soil using a peg if necessary.

Most shrubs root in less than a year, and if layered during the autumn or winter are often ready for transplanting the following autumn. A layered shoot should, however, be carefully examined before severing and, if insufficiently rooted, should be left for another year.

Practically any woody plant may be increased by layering, but the following genera are particularly suitable: magnolia, pieris, rhododendron, syringa (lilacs), plagianthus, garrya, photinia, mespilus, leucothoë and nothofagus. Certain procumbent or horizontally growing conifers often seen in rock gardens are convenient plants to layer. Various other dwarf shrubs found in the rock garden may be layered *in situ*. These include *Daphne cneorum* and dwarf rhododendrons.

Tip layering

Perhaps the simplest form of layering is tip layering or growing point layering, which often occurs naturally and is the normal method of increasing blackberries and loganberries (Fig. 62). It may also be used for currants and gooseberries. Tip

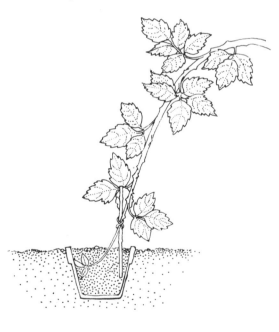

Fig. 62. Tip layering. A small number of woody plants will root if the growing tip is buried under the ground. The roots will occur at a leaf junction and, having rooted, the tip will then grow on upwards into the light forming a new plant.

layering consists of bending down firm young shoots in summer and burying their tips in the soil to a depth of 10 to 15 cm (4 to 6 in) or they may be inserted in pots. Rooting usually occurs in a matter of weeks and the new plants may be severed and transplanted in the autumn.

Serpentine layering

Serpentine layering is also easy to perform, but is only suitable for plants with slender pliable stems, such as honeysuckle, jasmine and lapagerias. Stems of these are laid in the ground and several parts of them layered, alternating with unlayered portions or curves whose buds provide shoots for the new plants. Sometimes the shoots are layered into pots sunk in the ground.

Continuous layering

Another method is called continuous layering (Fig. 63) and is popular for increasing fruit tree root stocks. First of all it is necessary to establish a permanent row of plants, each with a single stem. These stems are pegged down on the ground to form a continuous line of layered shoots in each row. The young shoots that arise from these stems are gradually earthed up as they grow, to a depth of 15 to 20 cm (6 to 8 in). Layering of this type is also done in the autumn or winter, and at the end of the growing season rooted shoots are removed and transplanted to a nursery bed. At least one young shoot should be left near the base of each stock for layering again the following year.

This form of layering is sometimes called the etiolation method because the bases of the young shoots are blanched or etiolated by the soil covering. As explained previously, etiolation promotes rooting and sometimes stems are pretreated in this way before being made into cuttings. As well as fruit tree root stocks, certain shrubs with long slender stems may be increased by continuous layering. Examples are *Cotinus coggygria*, *Hydrangea paniculata* and *H. arborescens*.

Mound layering

Mound or stool layering (Fig. 64) is somewhat similar to continuous layering. In this case the plants may be set closer together as their stems are not pegged down. Instead they are cut off at about ground level. Young shoots spring up from the base and as they grow they are earthed up. Fruit tree rootstocks are usually mounded up to a

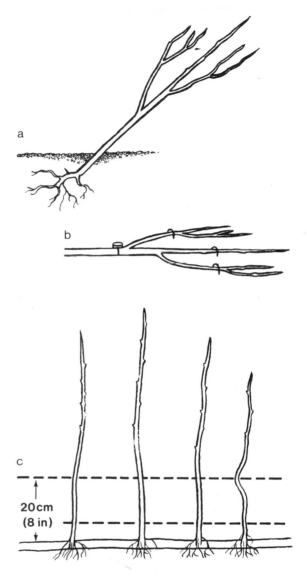

Fig. 63. Continuous or etiolation layering. (*a*) The young root-stocks are planted 60 cm (2 ft) apart at an angle of 45 degrees during the winter. (*b*) The following winter they are pegged down to the soil in a trench 5 cm (2 in) deep. Cover with 2.5 cm (1 in) of soil. (*c*) As the shoots grow they are earthed up gradually like potatoes. The lower dotted line shows the height of early earthing up and the upper dotted line the final depth of the earthed up soil. The rooted shoots can be removed and transplanted the following winter, and a few of them left behind to continue the layering process.

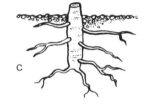

Fig. 64. Mound layering or stooling. (*a*) The root-stocks are planted 30 cm (1 ft) apart in winter and immediately cut back to within 45 cm (18 in) of ground level. (*b*) and (*c*) They are allowed to grow freely in the following year, but during the next winter are cut to within 2.5 cm (1 in) of the ground. (*d*) When shoots appear in spring they are left to grow to a height of about 15 cm (6 in), before earth is drawn around them. Eventually the soil should be mounded up to a depth of 15–20 cm (6–8 in). (*e*) The rooted stems may be dug up, severed and replanted the following winter.

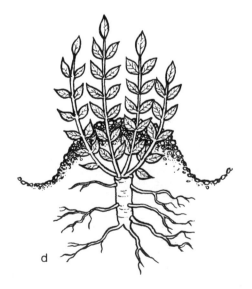

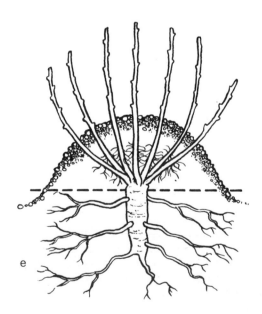

depth of 15 to 20 cm (6 to 8 in) but a thin layer of soil is sufficient in most other cases. Blackcurrants and gooseberries are very easily propagated in this way as well as a number of ornamental shrubs. It is particularly appropriate for those with stiff upright stems, such as deutzias and philadelphus.

A special type of layering (sometimes known as 'dropping') is adopted for heaths. Strong healthy plants 3 to 4 years old are best for this purpose. These are lifted and transplanted so that only 2.5 cm (1 in) or so of their tips is above soil level. The new planting hole should be filled in with a mixture of soil, sand and peat, and the compost should be worked carefully amongst the branches and shoots and made firm. This type of layering induces root production from the shoots, and if the work is done in early autumn they are ready for lifting the following autumn. If preferred, transplanting may be delayed until spring. Lifting is done with a sharp spade which severs the rooted portions of the stems from the parent plant beneath. Several of these stems are bunched together and treated as a single plant. They may be set in nursery beds or in their permanent positions.

Certain dwarf shrubs, such as ledums, gaultherias and vacciniums, are layered by 'dropping' them slantwise into the soil so that the bases of the young shoots near the tips of the spread-out branches are buried in the soil (Fig. 65).

Herbaceous layering

The layering of border carnations is usually regarded as a skilled operation, but it is not really difficult. The work should be done in July or August when the plants are growing freely. The first step is to place a sandy compost around the plants to be layered. Before starting to layer, some pieces of wire bent in the shape of a hairpin should be secured. Select young shoots and remove a few of their lower leaves. Next make a slit near the base about half-way through the stem and in an upward direction. The stem is then bent to open the slit and at this point is pegged firmly into the soil and should be covered with 2.5 cm (1 in) or so of the compost. It is usually possible to layer several stems on each plant.

The compost around the plants should be kept moist, and watering with a fine rose may be necessary. The layers will be rooted by the autumn and should then be severed and potted.

Fig. 65. Heathers can easily be layered by taking out the soil around them in the form of a shallow bowl and filling up the centre with soil, while keeping the shoots pressed back against the sides of the bowl.

The pots are usually stood in a cold frame over the winter and are ready for planting out in the spring.

A number of herbaceous plants may be layered, particularly alpines that are difficult to strike from cuttings; an example is *Acantholimon glumaceum*. The method is to select a young shoot and press it back as if taking a heel cutting. It is only, however, severed about halfway, and is then pegged down on sandy compost. Another form of layering is simply to work in some light compost among the stems of tufted alpines such as many campanulas. This induces rooting at the stem bases which may then be pulled off and potted. This method provides an alternative to cuttings and can also be used prior to division. It is found that a plant treated in this way will usually divide into a greater number of plants than one untouched.

103

Air layering

In all the various forms of layering described so far, rooting is induced by bringing the stems into contact with the soil or other moist medium. There is one quite distinct type of layering whereby the moist material is brought to the stem, the latter being left in its original position. This is called 'air layering', 'Chinese layering' or 'marcotting', and is believed to have been used by gardeners for thousands of years. It is a useful method for propagating rare and valuable plants which are difficult or impossible from cuttings, and where ordinary methods of layering are not practicable. Plants with rigid branches borne high above the ground could not be brought down to the soil so the soil is taken up to them.

Clean young shoots are best for air layering. They are prepared by making an upward cut about 5 cm (2 in) long at or about the stem centre. Alternatively, the stem may be girdled by removing a ring of bark about 1 cm ($\frac{1}{2}$in) wide. The wound may be dusted with a growth-promoting substance, and a handful of moist sphagnum moss is packed between the cut surfaces and all around it to give complete cover (Plate 4).

In the past, the main difficulty with air layering was to prevent the moss from drying out. This problem has now been overcome by the use of polythene film. This is wrapped around the sphagnum moss in such a manner as to leave no opening which would allow evaporation of moisture from the moss. The film is held in position by a piece of sticky tape around the stem both above and below the layer.

The advantage of polythene film is that it allows the transfer of air but practically no moisture can escape from inside it. This has been proved in American experiments where, with air layers in the open, the moss remained damp for as long as twelve months when covered with polythene film. These experiments also showed that it is important to prevent rain seeping into the moss. A wet soggy condition of this material may prevent rooting.

Air layering may be done in the spring or during the summer. Rooting often occurs within a few months, and is rarely delayed longer than a year. Considerable care is necessary with the new plants after they have seen severed from their parent. As a rule they are potted, using an open sandy compost. The pots should be stood in a shaded position such as a cold frame provided with a shaded light (if the plants are hardy) or a shaded corner of the greenhouse, until they are well established and showing signs of growth. Careless handling in the transfer from moss to soil may kill the young roots.

A wide range of plants may be increased by air layering, including difficult species such as acers, cornus, hibiscus, and wisteria. The rubber plant, *Ficus elastica*, can be easily propagated in this way when it grows too tall or sheds its lower leaves.

Pot layering

This old world method is still suitable for pliable slender-stemmed greenhouse climbers, otherwise difficult to root. Examples are bougainvillea, bignonia, lapageria, mutisia and stephanotis. The technique involves the use of two pots of a size convenient for the purpose, but one pot being larger than the other.

Begin by placing the smaller pot inside the larger one. Then stand both pots adjacent to the plant to be layered. Then thread one or more shoots through the drainage hole in the large pot, bringing each stem upwards through the space between the pots and out between rims. Carefully holding the stems in position fill in this space with light sandy compost so that the shoots are firmly embedded in the compost. The centre pot should also be filled with soil or compost to retard drying out. As with other methods of layering some form of stem wounding such as cutting a slice off one side of the stem promotes rooting.

This method of propagation is rarely used today, but greenhouse gardeners may be interested to try it. It has been used successfully to propagate *Carpenteria californica*.

14
The Propagation of Trees and Shrubs

Seed is widely used in the propagation of trees and shrubs. Autumn sowing immediately the seed is ripe is often advantageous. However, as the seed is unlikely to germinate before late spring or early summer weeds may be a problem. With such seed, therefore, over winter stratification may be preferable. A simple method of doing this is to mix the seed with moist soil, peat or sand or a mixture of these and place in boxes or pots which are buried outside.

In the propagation of shrubs from softwood or semi-hardwood cuttings mist can be used almost without exception.

How the different kinds are increased

ABELIA In July take half-ripe sideshoots with a heel and insert in a propagating frame. Mature sideshoots will also strike if planted in a cold frame in November.

ABIES (silver fir) Sow in the open choosing a shady position in spring. Transplant the seedlings in the autumn. Rare species should be sown in a heated greenhouse. Scions 5 to 8 cm (2 to 3 in) long, made from leading shoots only, may be grafted on seedling stocks using the whip and tongue method.

ABUTILON (Indian mallow) Softwood cuttings root readily in June or July in a propagating frame. Hardwood cuttings may be struck in a cold frame in November. *A. vitifolium* may be raised from seed sown under glass in January.

ACACIA (mimosa; wattle) Several of the species may be increased from seed in gentle heat. Cuttings of half-ripe wood taken with a heel will strike with bottom heat.

ACER (maple) Most of the species and certain varieties, such as *A. palmatum* 'Atropurpureum' may be raised from seed sown in the open as soon as ripe or stratified and then sown. Varieties of *A. negundo* and *A. palmatum* may be budded on the seedling species. The latter is also grafted under glass.

ACTINIDIA Cuttings of half-ripe wood will root in moderate heat. Seed in gentle heat is another method while shoots may be layered during the dormant season.

AESCULUS (horse chestnut) The species of the horse chestnut are usually increased by seed, sown when ripe 8 cm (3 in) apart and 1 cm ($\frac{1}{2}$ in) deep. Budding in summer or grafting in April on seedling stocks are other methods. The common horse chestnut should not, however, be used as a stock as one of the other species gives better results. *A. parviflora* also grows readily from root cuttings 7 to 8 cm ($2\frac{1}{2}$ to 3 in) long planted outside in March or April.

AILANTHUS (tree of heaven) Easily raised from seed sown outdoors when ripe, but to secure plants of the female type only, which are preferable, take cuttings 8 cm (3 in) long from female plants and plant outside in spring.

AKEBIA Cuttings from half-ripe wood will root with a little heat. Layering in early spring is another method.

ALBIZIA Increased by seed sown in a warm greenhouse in spring.

ALNUS (alder) Seed collected in the autumn is sown outside in spring. Stratifying often improves germination. Varieties may be grafted on the common alder *A. glutinosa* under glass, or hardwood cuttings rooted outdoors after leaf fall.

AMELANCHIER The species of this genus are increased from seed gathered when ripe then stratified for 18 months and sown in the open in spring. Layering the wood of the previous season's growth is another method. Some species may be propagated from suckers.

AMPELOPSIS Softwood cuttings can be struck in a propagating frame in July, and in October hardwood cuttings will root in a cold frame.

ANDROMEDA (marsh rosemary) Sow seed in early spring under glass in a compost consisting mainly of sieved moist peat and a little sand. Layering may be carried out in August in a shady situation. Division of plants previously 'dropped' in peaty soil is effective. From June until August half-ripe sideshoots taken with a heel will strike in a mixture of peat and sand in a propagating frame.

ANTHYLLIS (lady's fingers) Shrubs in this genus may be raised from seed or cuttings in a cold frame in spring.

ARALIA (angelica tree) Usually increased from suckers and root-cuttings in spring. Seed is another method.

ARAUCARIA Seed of *A. araucana* (monkey puzzle) can be sown in early spring under glass. *A. heterophylla* (syn. *A. excelsa*), the Norfolk island pine, is raised from cuttings of leading shoots. The procedure is to remove the leading shoot from the parent plant and use it as a cutting. Buds are then produced in the axils of the upper branches and these develop into leading shoots which are again used as cuttings, and at the same time the tree is cut back to the next tier, where a further lot of new leaders is induced to grow. Continue in this way until the lowest tier is reached and the tree is then discarded. Cuttings are made from semi-mature wood of the current season's growth secured in June and July, and should be inserted in a sandy compost contained in small pots. Place these in a propagating frame provided with bottom heat.

ARBUTUS *A. unedo* (the strawberry tree) grows readily from seed secured from ripe fruits. Varieties are grafted on seedling stocks under glass in early spring. Ripe cuttings planted in a cold frame may be tried, but results are often poor. Layering is also possible but usually difficult owing to the type of growth.

ARCTOSTAPHYLOS (bearberry) Seed sown in spring or softwood tip cuttings are the usual methods; also layering.

ARISTOLOCHIA (Dutchman's pipe) Softwood cuttings in a cold frame in summer will root but bottom heat produces better results.

ARONIA (chokeberry) Propagation is effected by seed sown in autumn, by late summer cuttings or division.

ARUNDINARIA (bamboo) Divide old plants in spring and for speedy rooting pot the divisions and keep in a warm, humid greenhouse.

ATRIPLEX Propagate by summer cuttings.

AUCUBA (spotted laurel) Cuttings made from firm shoots 15 to 23 cm (6 to 9 in) long and planted in the open during the autumn usually strike readily. Layering is a still more certain method and can be recommended to the amateur.

AZARA Insert cuttings made from mature sideshoots in a cold frame in autumn. The variegated type is usually layered.

BALLOTA Cuttings of young shoots can be rooted in a garden frame in summer.

BERBERIDOPSIS (coral berry) May be raised from spring-sown seed, or cuttings of young shoots may be rooted in gentle heat.

BERBERIS (barberry) Many species come true from seed including *B. darwinii*, *B. gagnepainii* and *B. thunbergii* if each is reasonably isolated from other species. Stratify the seed over the winter and sow in the open in early spring. Special varieties and hybrids such as *B.* × *stenophylla* do not come true from seed and must be propagated vegetatively. Cuttings of a large number of both evergreen and deciduous types are made about 15 cm (6 in) long from current season's wood with a heel, and strike easily in a cold frame. Layering and division are alternative propagation methods.

BETULA (birch) The seed is sown when ripe in soil made light with peat and sand. Cover very lightly and give some protection during the winter with brushwood or bracken. Varieties of the silver birch *B. pendula* are grafted on to seedlings of the ordinary type in February under glass. Budding is sometimes attempted on established stocks in the open in July.

BUDDLEIA Cuttings are a popular and easy method and will strike in the open. Use hard wood and make the cuttings 15 to 20 cm (6 to 8 in) long. Plant in October or November. Softwood cuttings 8 to 10 cm (3 to 4 in) long also strike readily in a propagating frame in July. *B. alternifolia* is often raised from seed sown in heat in February.

BUPLEURUM Cuttings can be rooted in a propagator in late summer.

BUXUS (box) Most of the boxwoods are easy from cuttings. Take mature sideshoots with a heel and plant in a cold frame in September. Division is another method.

CALLISTEMON (bottle-brush tree) Increased by cuttings of firm young shoots inserted in a propagating frame in summer or from seed.

CALLICARPA (French mulberry) Seed and softwood cuttings in heat are the usual methods.

CALLUNA (ling) Propagate as for ericas.

CAMELLIA C. japonica is raised from seed sown in heat in February. Prior to sowing it is recommended to soak the seed in warm water for 24 hours. C. japonica and C. reticulata varieties are usually grafted on seedlings of the ordinary type. This is done in a propagating frame in August. Several varieties may be raised from cuttings of half-ripe wood in a propagating frame in July. Another method is leaf-cuttings. (See Chapters 9 and 11.)

CAMPSIS Increase by seed if available. Other methods are hardwood cuttings, suckers and root cuttings in moderate heat.

CARAGANA Cuttings may be rooted in a frame in summer; seeds may be sown in a frame in spring after being soaked in hot water. Certain species may be grafted on to stocks of C. arborescens in spring.

CARPENTERIA (Californian mock orange) Seed of C. californica (the only species) is sown in heat in February but does not usually germinate freely. Soft young sideshoots taken with a heel will root in propagating frame in summer.

CARPINUS (hornbeam) The species are raised from seed sown in open beds in the autumn. Rare forms and varieties are grafted on seedlings of C. betulus (common hornbeam). This is done in early February under glass.

CARYOPTERIS Small softwood cuttings can be secured from plants grown in heat in March and are struck in a propagating frame. Cuttings of a similar type secured from outdoor plants in July may be similarly rooted. Hardwood cuttings 10 to 15 cm(4 to 6 in) long planted in a cold frame in November provide another method.

CASSINIA The most usual method is to take cuttings 10 to 15 cm (4 to 6 in) long with a heel in August or September and plant in a cold frame.

CASSIOPE Small cuttings taken with a heel are inserted in a cold frame in August. Use a peaty compost and shade from strong sun. C. hypnoides is increased by the division of 'dropped' plants.

CASTANEA (Spanish chestnut) C. sativa is increased by seed collected when ripe. Stratify for a few months and sow outside in February. Varieties are grafted in April or budded in July on seedling stocks.

CATALPA (Indian bean tree) Seed of the species is sown outdoors in April in a sheltered position. Root cuttings about 4 cm (1½ in) long are inserted horizontally in boxes or singly in small pots. Cover with sand and place in a propagating frame. Cuttings of firm young growths can be rooted in a propagating frame in summer.

CEANOTHUS Sideshoots with a heel are taken in October and rooted in a cold frame. Softwood cuttings will also strike in June in a propagating frame with bottom heat. The evergreen species such as C. rigidus are best propagated from mature sideshoots inserted in a cold frame in autumn.

CEDRUS (cedar) The species are best raised from seed sown outside in early spring. Varieties and special forms are grafted on seedlings of the same species.

CELASTRUS (staff vine) Layer young shoots in spring or autumn.

CERATOSTIGMA (leadwort) Softwood cuttings with a heel strike readily in a propagating frame. These are secured from outdoor plants in summer or from plants grown in pots which are cut back in the autumn and brought into heat in December or January to force into growth.

CERCIS The species are raised from seed sown in heat in February or March. The varieties of C. siliquastrum (Judas tree) are grafted in winter under glass on to seedling stocks grown in pots.

CESTRUM (bastard jasmine) Increase by half-ripe cuttings with a heel in a propagating frame with bottom heat.

CHAENOMELES The well-known C. lagenaria (syn. Cydonia japonica) (Japanese quince) and other species may be raised from seed which is collected when ripe, stored dry over winter and sown in early spring. Varieties are often grafted on to seedling stocks, but plants on their own roots are superior and may be had by layering

CHAMAECYPARIS (false cypress) The ordinary species such as C. lawsoniana (Lawson's cypress) are raised from seed sown in the open in spring. Special forms and varieties are raised from cuttings consisting of sideshoots with a heel. These are inserted in a cold frame in October. Grafting on seedling stocks under glass is also used to increase the varieties.

CHIMONANTHUS (winter sweet) Sow seed of *C. praecox* in early spring. Layering is another method often used.

CHIONANTHUS (fringe tree) Seed may be sown outdoors or under glass after being stratified. Cuttings of firm young growths may be rooted in a cold frame in summer.

CHOISYA (Mexican orange blossom) *C. ternata* may be increased by cuttings of mature wood 10 to 15 cm (4 to 6 in) long with a heel rooted in a cold frame in October. The immature tips root readily in a propagating frame and are taken in July.

CISTUS (sun rose) True species are raised from seed sown under glass in February. Cuttings of sideshoots taken with a heel root readily in a propagating frame in summer. Wood that is almost mature may be used for cuttings in October which are planted in a cold frame.

CLEMATIS Seed may be used when available to increase the species. For varieties internodal cuttings of young growths are rooted in a propagating frame in summer. The lower cut is made between two leaf joints and the upper cut above a pair of leaves. Only one pair of leaves should be present on each cutting. It is advantageous to cut a thin slice of bark off one side of the cutting from about 2.5 cm (1 in) above the base, downwards. Grafting is still used occasionally for the various hybrids. *C. vitalba* (traveller's joy) is used as a stock, the scions being whip grafted on to the roots under glass.

CLERODENDRUM Usually propagated from root cuttings. Fairly thick pieces about 8 cm (3 in) long are secured in winter and may be laid in moist sand until April when they are planted in outdoor beds.

CLETHRA (white alder) Increase is by seed sown in early spring; summer cuttings, and spring or autumn layering.

CLIANTHUS (glory pea) *C. formosus*, (*syn. C. dampieri*) is grafted in the seedling stage on seedlings of *Colutea arborescens*. *C. puniceus* is raised from cuttings of half-ripe wood about 8 cm (3 in) long inserted in a propagating frame June–July.

COLLETIA Cuttings of firm young growths can be rooted in a frame in summer or autumn.

COLUTEA (Bladder senna) The best plants are produced from seed which is kept dry until sowing time in February or March under glass. Half-ripe nodal cuttings may be rooted in a propagating frame in July or August.

CONVOLVULUS *C. cneorum* and *C. mauritanicus* can be increased by cuttings of firm young growths inserted in a propagating frame in summer.

CORNUS (dogwood) Several species such as *C. alba* and *C. nuttallii* are raised from seed sown under glass in March. Cuttings of half-ripened sideshoots may be struck in a propagating frame in summer. Mature sideshoots with a heel will root in a cold frame if inserted in autumn. Layering is, however, the most generally used method and, if done in July, the layers will be ready for lifting the following spring.

COROKIA May be raised by seed in spring or by cuttings in July or August.

CORONILLA (crown vetch) Easily raised from spring-sown seed or summer cuttings.

CORYLOPSIS Layering is the usual method but cuttings in late summer will also strike.

CORYLUS (hazel) Layering in the autumn is the most usual method. Suckers are produced by some species and are usually transplanted in spring. The varieties of *C. avellana*, such as 'Contorta' and 'Pendula' are grafted under glass in winter or budded in summer, the stock being the true species in each case. 'Contorta' may also be propagated from hardwood cuttings.

COTINUS Suitably pendent stems can be layered in spring; cuttings of firm young growths can be rooted in a propagator in summer.

COTONEASTER Seed is successful for many species including *C. bullatus*, *C. frigidus*, *C. lacteus* and *C. simonsii*. It should be collected when ripe, stratified during the winter and sown in the open in March. Cuttings are used for several species, such as *C. adpressa*, *C. horizontalis*, *C. microphyllus* and *C. prostratus*. These should always be taken with a heel and may be either ripe wood taken in November and planted in a cold frame, or half-mature sideshoots inserted in a frame in July.

CRATAEGUS *C. monogyna* (common hawthorn) is readily raised from seed. The haws are gathered when ripe and mixed with moist sand and left outside for 18 months to stratify the seed. This is then sown in beds in the spring. A large number of other species are also raised from seed. The various ornamental hawthorns are budded in summer or grafted in spring on the common hawthorn.

CRYPTOMERIA This is raised by seed sown in the open in spring. Cuttings are also used for *C. japonica* (Japanese cedar) and its varieties, and are taken with a heel in early autumn and planted in a cold frame.

× CUPRESSOCYPARIS (Leyland cypress) Cuttings of firm young growths can be rooted in a propagator or under mist in summer.

CUNNINGHAMIA Sow seeds in a temperature of 18°C (65°F) in spring.

CUPRESSUS (cypress) *C. macrocarpa* and *C. sempervirens* are raised from seed sown in February or March indoors, or outside later in the spring. Varieties may be grafted under glass in summer, *C. macrocarpa* being used as a stock.

CYTISUS (broom) Several species are easily raised from seed, but crossing occurs readily unless the plants are well isolated. Sow in March or April using pans placed in a cold frame. Cuttings are used for some species and several varieties, e.g. *C. ardoinii*, *C. × praecox*, *C. purgans* and *C. purpureus* and its varieties. These may be either half-ripe sideshoots with a heel inserted in a propagating frame in August, or firm sideshoots secured in the autumn and planted in a cold frame. Besides cuttings, some types such as *C. × kewensis* and *C. × beanii* can be grafted on to seedling laburnum in March, in the open.

DABOECIA *D. cantabrica* (syn. *D. polifolia*, Irish heath) may be raised from seed sown in spring under glass in pure sifted peat. Cuttings may be struck in a cold frame in October. Division of plants previously 'dropped' is very effective on light, sandy soils.

DAPHNE Seed is used for several species such as *D. laureola* (spurge laurel) and *D. mezereum* (mezereon) and its varieties. Stratify the seed during the winter and sow in early spring in a cold frame. Cuttings are successful for *D. cneorum* (garland flower), *D. collina* and *D. retusa*. They consist of half-ripe sideshoots inserted in a cold frame in summer. Root cuttings in heat is the usual method for *D. genkwa*. Another method is grafting under glass in spring, *D. laureola* being the stock for evergreen species and *D. mezereum* for deciduous types.

DESFONTAINEA Seed of *D. spinosa* is sown under glass in early spring. Cuttings of half-ripe shoots with a heel are inserted in a propagating frame in early July. Firm sideshoots may be rooted in October in a cold frame.

DEUTZIA All species and varieties are readily increased by softwood cuttings 10 to 12 cm (4 to 5 in) long taken in summer and inserted in a propagating frame. Hardwood cuttings about 20 cm (8 in) long are usually taken with a heel and in mild districts can be planted outside in November. In colder localities they should be inserted in a cold frame and made somewhat shorter. Seed stored dry over winter will germinate in a warm greenhouse in February.

DIERVILLA (bush honeysuckle) Softwood cuttings, 10 to 12 cm (4 to 5 in) long with or without a heel can be rooted in a propagating frame in July. Ripened wood cuttings about 15 cm (6 in) long may be planted in a cold frame in November or in a sheltered spot in the open. Suckers may be removed and transplanted in autumn or winter.

DRIMYS Suitably pendent stems can be layered in spring; cuttings can be rooted in a propagator in summer or autumn.

ECCREMOCARPUS *E. scaber* (Chilean glory flower) is easily raised from seed, which is abundantly produced. Sow in pans and leave outside over winter. In February bring indoors for quick germination.

ELAEAGNUS Deciduous species can be raised from seed stratified about 18 months and sown in early spring in heat. Softwood cuttings of *E. multiflora* and *E. glabra* will root in a propagating frame. Firm cuttings 10 to 12 cm (4 to 5 in) long with a heel can be struck in a cold frame in autumn. Layering is approved for several difficult species such as *E. angustifolia* (oleaster), *E. macrophylla* and *E. umbellata*.

EMBOTHRIUM Grown from seed sown in spring in brisk heat, stem or root cuttings in a propagating frame and by grafting young shoots on to seedling stocks in spring.

ENKIANTHUS Normally raised from seed collected when ripe in winter, stored dry and sown in March in a peaty compost. Layering involves pegging down the young shoots in spring. Cuttings of firm young growths may be rooted in a propagator in summer.

ERICA (heath) The usual method is by cuttings of semi-mature tips about 3 cm (1½ in) long. Insert these in a compost of peat and sand in small pots and place in a propagating frame. Firm cuttings can also be rooted similarly if taken in October–November. Layering or 'dropping' is another method and many species can be easily divided in spring. Seed is used to propagate a few species such as *E. arborea* (tree heath) and *E. lusitanica*

(Portuguese heath). It should be sown on sifted peat in February.

ERYTHRINA Heel cuttings can be rooted in a propagator in spring; seeds can be sown in a temperature of 18°C (65°F) in spring.

ESCALLONIA Softwood sideshoots 5 to 10 cm (2 to 4 in) long root in a propagating frame, and mature shoots of a similar type can be inserted in a cold frame in October, or even outside in a sheltered position.

EUCALYPTUS (gum tree) Easily raised from spring-sown seed in moderate heat.

EUCRYPHIA Increased by layers or late summer cuttings inserted in a propagating frame. Seeds may be sown in peaty soil in spring.

EUONYMUS Softwood cuttings of *E. japonicus* with or without a heel strike in a propagating frame in summer. Cuttings of ripened wood with a heel, planted in October, can be rooted in a cold frame. *E. radicans* can be rooted from softwood cuttings. Division is suitable for some varieties and several species can be raised from seed sown in spring.

EURYOPS Cuttings can be rooted in a propagator in summer; seeds can be sown in a temperature of 16°C (60°F) in spring.

EVODIA Sow seed in early spring.

EXOCHORDA The species are usually raised from seed collected when ripe, stored dry over winter and sown in February. Cuttings of young shoots may be rooted in a propagating frame in summer.

FABIANA Cuttings of young shoots may be rooted in a propagating frame in summer.

FAGUS (beech) *F. sylvatica* (common beech) is raised from stratified seed sown outside in spring. Rare species should be sown indoors. Varieties of *F. sylvatica* are grafted on to the species either in the open in April or under glass in winter.

FATSIA (false castor oil) Usually raised from seed, but cuttings of firm young growths may be rooted in a propagating frame in summer.

FORSYTHIA *F. suspensa* (golden bells) Hardwood cuttings 15 to 20 cm (6 to 8 in) long may be planted in a cold frame in October or in a sheltered site in the open. Semi-mature tips root readily in a propagating frame in summer.

FOTHERGILLA (American witch hazel) Increased by layering in autumn or spring. Cuttings of half-ripe wood will strike in late summer in moderate heat.

FRANCOA (bridal wreath) Divide plants in spring; sow seeds in a temperature of 16°C (60°F) in spring.

FRAXINUS *F. excelsior* (common ash) is raised from seed sown when ripe in open beds. The weeping ash is increased by grafting on to seedlings of the common type.

FREMONTODENDRON Raised from spring-sown seed in moderate heat. Cuttings of young growths may be rooted in a propagator in summer.

FUCHSIA Cuttings made from soft tips when available can be rooted in a propagating frame at any time of the year.

GARRYA Layering in the autumn is the usual method. Firm sideshoots 8 to 10 cm (3 to 4 in) long with a heel will root when inserted in a cold frame in October. Half-ripe shoots can be rooted in a propagator in summer.

GAULTHERIA Several species including *G. shallon* are raised from seed sown in February. Layering and division are other suitable methods. Certain species such as *G. hispida* and *G. oppositifolia* can be raised from cuttings secured in August and planted in a frame.

GENISTA *G. hispanica* (Spanish gorse) and other species are raised from cuttings of ripe sideshoots 5 to 10 cm (2 to 4 in) long planted in a cold frame in September. Several species may be raised from seed sown under glass in February.

GINKGO (maidenhair tree) *G. biloba* is usually raised from seed stratified over winter and sown in the open in March. Varieties are grafted on to the type under glass. Cuttings of firm young growths may root in a propagator in summer.

GRISELINIA Readily increased by sideshoots with a heel inserted in a cold frame in October, or softwood cuttings in a propagating frame in summer.

HALESIA (snowdrop tree) *H. carolina* is propagated by spring-sown seed or by autumn layers.

HAMAMELIS (witch hazel) In nurseries grafting under glass in August is the usual method of increase. The stock used is *H. virginiana* which is raised from seed sown in a cool greenhouse. Layering in spring is the most suitable method for the amateur.

HEBE Seeds of true species may be sown under glass in spring. Cuttings of firm young growths may be rooted in a propagator in summer; riper shoots in a cold frame in autumn.

HEDERA (ivy) Ivy is normally increased from cuttings, either the softwood type taken in summer and inserted in a propagating frame, or firm nodal cuttings planted in a cold frame in October. The shoots can also be layered at any time of year.

HIBISCUS (rose mallow) The shrubby species of

hibiscus are usually propagated from hardwood cuttings in early autumn with bottom heat. Layering and autumn-sown seeds are other methods. Grafting on seedling stocks in heat in January is also successful.

HIPPOPHAE (sea buckthorn) Increased by root cuttings outdoors in spring; by autumn-sown seed, and by layering.

HOHERIA Cuttings of firm shoots can be rooted in a propagator in summer; seeds can be sown in a temperature of 16°C (60°F) in spring.

HYDRANGEA *H. macrophylla* (syn. *H. hortensia*), and its varieties are best propagated from softwood cuttings taken from plants grown indoors or outdoors. The cuttings are potted singly into small pots, using sandy compost. The pots are placed in a propagating frame. *H. paniculata* and *H. arborescens* may be mound layered in autumn. *H. petiolaris* is raised from seed sown in heat in spring.

HYPERICUM The species are raised from seed stored dry and sown in spring under glass. Cuttings are also generally used (either of firm wood with a heel inserted in a cold frame in autumn, or softwood cuttings in a propagating frame in summer). *H. calycinum* (rose of Sharon) is easily increased by division.

ILEX (holly) The common holly and other species are raised from seed stratified for 18 months and sown outside in spring. Layering in the autumn is another method. For this purpose stock plants must be partly lifted and laid on their sides, and each young shoot should be tongued before being layered. Budding of varieties on the type plant is also done. Half-ripe cuttings may be rooted under mist in summer.

JASMINUM All species and varieties are raised from cuttings. Hardwood cuttings with a heel are taken in November and inserted in a cold frame, or in a sheltered position in the open. Immature sideshoots also root readily in a propagating frame in summer.

JUGLANS (walnut) The species are raised from seed stratified over winter and sown in the open in early spring. Grafting is used for varieties but is difficult.

JUNIPERUS (juniper) Seed is used for several species. It should be stratified for about 18 months and then sown in the open in spring. Layering is a convenient method for the procumbent types and is carried out in autumn, each shoot for layering being first partly severed.

Many varieties grow from cuttings—sideshoots secured from August to October and planted in a cold frame.

KALMIA Seed is sown in peaty compost under glass in March. Layering is the best method for increasing varieties. It should be carried out in autumn, the layered shoots being twisted or tongued. *K. polifolia* and several other species will grow from half-ripe cuttings taken in August and inserted in a propagating frame.

KERRIA (Jew's mallow) Cuttings of firm shoots with a heel root when planted in the open in autumn. Softwood tip cuttings will also strike in a propagating frame in July. Suckers may be removed and transplanted in autumn or winter.

KOLKWITZIA Take half-ripe cuttings of *K. amabilis* (beauty bush) in July and insert in a propagating frame. Suitable stems may be layered in autumn or spring.

LABURNUM *L. anagyroides* (common laburnum, golden chain) is easily raised from seed sown in the open in spring. Hardwood cuttings 23 to 30 cm (9 to 12 in) long with a heel may be used to increase all varieties. Varieties may also be budded or grafted on common laburnum.

LAGERSTROEMIA Cuttings of firm young growths can be rooted in a propagator in summer; seeds can be sown in a temperature of 18°C (65°F) in spring.

LARIX (larch) Dried seed is sown in the open ground in spring. Varieties are grafted on seedlings under glass in summer.

LAURUS *L. nobilis* (sweet bay) is normally raised from cuttings of firm shoots with a heel planted in a cold frame in early winter. Layering is another method sometimes used.

LAVANDULA (lavender) Cuttings can be rooted outdoors in late summer or early autumn.

LAVATERA (tree mallow) Shrubby species are increased by seed sown in gentle heat in spring. Summer cuttings will also strike in a propagator.

LEDUM (marsh rosemary) Methods of propagation are seed in spring, summer cuttings, layering and division in autumn.

LEIOPHYLLUM (sand myrtle) Seed may be sown in March in a cold frame or layering carried out in autumn.

LEPTOSPERMUM All species and varieties may be raised from cuttings of firm young growths rooted in a propagator in summer. Half-ripe cuttings 5 to 8 cm (2 to 3 in) long will also strike in a cold frame in August.

LEUCOTHOË Propagated from spring-sown seed in gentle heat, or autumn layering; also by division.

LEYCESTERIA Easily increased from summer cuttings in a shaded frame, also by spring-sown seed. Established clumps may be divided in autumn.

LIGUSTRUM The ordinary privet L. *vulgare* and also L. *ovalifolium* are very easily propagated from hardwood cuttings which are made 25 to 30 cm (10 to 12 in) long and inserted in the open in autumn and winter. Various other species and varieties are best increased from softwood cuttings about 8 cm (3 in) long with a heel, and planted in a propagating frame.

LIPPIA (lemon-scented verbena) Shrubby species are easily increased by summer cuttings rooted in a propagating frame.

LIQUIDAMBAR Seed sown in autumn may take two years to germinate. Alternatively layer in spring.

LIRIODENDRON (tulip tree) Imported seed of L. *tulipifera* should be stratified over winter and sown outdoors in spring. Varieties are grafted under glass on seedlings of the type.

LONICERA The climbing types are increased by ripe cuttings about 15 cm (6 in) long inserted in a cold frame or by softwood tips, 8 cm (3 in) long with or without a heel inserted in June in a propagating frame. The shrubby species are increased similarly. L. *nitida*, the popular evergreen hedging plant, is readily increased by autumn cuttings planted in the open.

LYCIUM Hardwood cuttings will strike in a cold frame in autumn. Suckers provide another method and seed is also used.

MAACKIA Seed in spring and root cuttings are the recommended methods of increase.

MAGNOLIA Layering is probably the best method and should be carried out in spring (Plate 5). Only young shoots should be layered after being slit to form a tongue. Many species can be raised from seed which should be sown when ripe in the open. Small quantities may be sown in boxes or pans, left outside until February and then brought indoors. Varieties are grafted under glass on seedling stocks.

MAHONIA Raised from seed in the same manner as berberis. Several species may be propagated from cuttings. These are taken in June or July, made about 15 cm (6 in) long and inserted in a propagating frame.

MALUS (apple) The flowering apples are usually grafted on to seedlings of M. *pumila* (wild crab); see also PYRUS.

MESPILUS (medlar) This is increased by layering in spring or autumn. Stratified seed is sown in spring and fruiting varieties are grafted on to *Pyrus communis* (pear) outdoors in spring.

METASEQUOIA (dawn redwood) Cuttings of firm young growths can be rooted in a propagator in summer; seeds can be sown in a frame in spring.

MORUS (mulberry) Easily increased from hardwood cuttings about 30 cm (12 in) long with a heel planted in the open in autumn. Semi-mature sideshoots also root in a propagating frame in July. Seeds may be sown after being stratified for three months.

MYRTUS (myrtle) Easily increased from cuttings made from firm young growths and inserted in a propagator in summer. Seed may be sown in autumn or spring.

NANDINA May be increased from seed if available. Cuttings may be taken in late summer and inserted in a propagating frame with bottom heat but they root rather slowly.

NEILLIA Seeds may be sown in autumn or spring and cuttings rooted in a propagator.

NOTHOFAGUS (southern beech) Some species such as N. *dombeyi* and N. *procera* may be propagated by half-ripe cuttings inserted in a propagator in summer, but layering in spring is more usual. Seed may be sown in spring.

OLEARIA The species and varieties are readily increased from softwood sideshoots removed in summer and rooted in a propagator, or from mature laterals with a heel secured in October and planted in a cold frame.

OSMANTHUS Half-ripe cuttings may be rooted in a propagator in summer. Cuttings of mature sideshoots with a heel taken in October will strike in a cold frame. Most species can be layered in autumn, the young shoots being well tongued before pegging down.

× OSMAREA Increased by summer cuttings rooted in a propagator.

OSTEOMELES Propagated by seed in spring or cuttings in summer.

OXYDENDRUM The usual method of increase is by seed sown in early spring in a cool greenhouse. Summer cuttings and layering are other propagation methods.

PACHYSANDRA Divide mature plants in spring or autumn; cuttings of firm young growths can be rooted in a propagator in summer.

PARROTIA Seed sown in early spring is the usual method and layering is another method used.

Cuttings of firm young growths may be rooted in a propagator in summer.

PARTHENOCISSUS Cuttings of firm young growths can be rooted in a propagator in summer; riper shoots may be inserted outdoors in autumn; suitably pendent stems may be layered in spring.

PASSIFLORA (passion flower) Increase is by seed or summer cuttings.

PAULOWNIA Normally raised from seed sown under glass in spring or autumn. Root cuttings may be taken in winter.

PERNETTYA Variable plants are produced from seed sown in spring in the open. Varieties are raised from cuttings of small sideshoots taken in July or August and inserted in a propagating frame. Division is an easy method for the amateur and should be carried out in spring. Young shoots may be twisted and layered in autumn.

PEROWSKIA This is increased from half-ripe shoots without a heel secured in July or August and inserted in a propagator. Root cuttings may be taken in winter and seeds sown under glass in spring.

PHELLODENDRON Increased from seed in spring or heeled cuttings in July.

PHILADELPHUS (mock orange) All the species are easily increased from cuttings of ripe shoots about 23 cm (9 in) long with a heel planted in the open in a sheltered position and a well-drained light soil. Softwood sideshoots 8 to 10 cm long taken in July will strike in a propagating frame.

PHILLYREA (mock privet) Readily increased from summer cuttings rooted in a propagator.

PHLOMIS *P. fruticosa* (Jerusalem sage) and other species will root from cuttings of ripe shoots about 8 cm (3 in) long when planted in a cold frame in October. Softwood nodal cuttings also strike readily in a propagating frame in summer. All may be raised from seed.

PHORMIUM (New Zealand flax) Divide large plants in spring; sow seeds in a temperature of 18°C (65°F) in spring.

PHOTINIA Cuttings of firm young growths may be rooted in a propagator in summer. Suitable stems may be layered in spring.

PHYGELIUS (Cape figwort) Semi-mature sideshoots with a slight heel will strike easily in a propagating frame in June or July. Established plants may be divided in spring.

PHYLLODOCE Propagated by seed in spring, cuttings of half-ripe shoots in summer and layering in spring.

PHYLLOSTACHYS (Whangee cane) Propagation is the same as that described for arundinaria.

PICEA (spruce) The species are propagated from seed stored dry and sown outside in spring. Varieties of *P. abies*, *P. glauca* and *P. orientalis* may be raised from cuttings of firm terminal shoots secured in August, potted singly into small pots, and placed in a propagating frame or under mist. Mature terminal shoots about 8 cm (3 in) long may also be rooted in a cold frame. Certain varieties are grafted on the parent species under glass in autumn.

PIERIS Cuttings of half-ripe sideshoots about 8 cm (3 in) long with a heel may be rooted in a propagating frame in summer. Autumn layering is another method.

PINUS (pine) Seed stored dry is sown in the open in spring on light soil. Small lots of seed and rare species should be sown in trays or pans under glass. Varieties are grafted on seedlings of the parent species.

PIPTANTHUS Sow seed in boxes or pans in heat in spring. Insert cuttings of ripe shoots in a frame in autumn.

PITTOSPORUM The species are raised from seed sown under glass in March in a peaty compost. Cuttings of semi-mature shoots with a slight heel will root in a propagator in July. Varieties are grafted on to seedlings of the parent species under glass in winter.

PLATANUS *P.* × *acerifolia* (London plane) is raised from hardwood cuttings about 25 cm (10 in) long with a heel planted in the open. Layering in the autumn is another method recommended for *P. orientalis* (oriental plane). Sow seeds outdoors in autumn.

POLYGONUM The popular climber *P. baldschuanicum* is propagated from hardwood cuttings 15 to 20 cm (6 to 8 in) long with a heel. These should be potted singly into 8 cm (3 in) pots which are stood on the greenhouse staging or in a cold frame.

POPULUS (poplar) Hardwood cuttings about 20 cm (8 in) long planted in a light sandy soil in autumn strike readily. Some species can be increased from suckers or root cuttings. Certain varieties must be grafted, and this is usually carried out in winter on the stock. *P. canescens* (grey poplar).

POTENTILLA The shrubby species such as *P. fruticosa* may be raised either from small softwood sideshoots taken with a heel and inserted in a propagating frame in July, or from hardwood sideshoots planted in a cold frame in October.

PRUNUS All the flowering cherries are budded or grafted on to various stocks in a similar manner to fruit trees. Budding is usually carried out in July or August and grafting in April in the open on established stocks. The stocks used include *P. avium* (sweet cherry) which is preferred for the Japanese cherries; *P. cerasifera* (myrobalan) is used for the varieties of this species; seedling peach for the flowering peaches and Brompton or St. Julien A for *P. triloba* and its varieties. Evergreen prunus species such as *P. laurocerasus* (cherry laurel) and *P. lusitanica* (Portugal laurel) may be increased by half-ripe cuttings rooted in a propagating frame in summer.

PSEUDOTSUGA This genus includes the Douglas fir which is propagated by sowing dried seed of the previous year in spring in the open.

PTEROSTYRAX Usually raised from seed when available.

PUNICA (pomegranate) Increased by cuttings of firm shoots inserted in summer in a propagating frame provided with bottom heat. Seed may be sown under glass in autumn or spring.

PYRACANTHA (firethorn) The species and varieties are increased by inserting small half-ripe sideshoots with a heel in a cold frame in August. Seeds are sown in a frame in spring after being stratified for three months.

PYRUS (pear) To produce pear seedlings, the pips should be stratified over winter and the seed sown in spring in the open or in trays in a cold frame or greenhouse. Other species and varieties may be grafted on to these seedlings in March in the open.

QUERCUS (oak) Acorns should be sown outside immediately they are collected, or they may be stratified over winter and sown either in the open or under glass. Varieties are usually grafted on to seedlings of the common oak under glass in winter.

RAPHIOLEPIS *R. indica* and *R. umbellata* are raised from seed sown under glass immediately it is ripe. Varieties are raised from half-ripe sideshoots with a heel taken in September and inserted in a propagator. *R. × delacourii* is grafted on seedlings of *R. umbellata* under glass in winter.

RHAMNUS Layer suitably pendent stems in spring or autumn; take cuttings of firm young growths in summer and root in a frame; sow seeds outdoors in autumn.

RHODODENDRON This very large genus is increased by various methods. Seed is used to raise *R. ponticum* (common rhododendron), which is widely used as a stock for grafting. The seed is collected when ripe and sown in the open in spring or in pans or trays in a cold frame. A very wide range of species is also increased by seed. Sow with care in pots or pans containing sifted peat. In nurseries a large number of varieties are saddle grafted, this being carried out in winter under glass. Layering is also widely used in nurseries, stock plants being established in peaty soil. Young shoots are layered in autumn (Plate 6) and before each is pegged down the stem should be well twisted. Cuttings are mainly used to propagate the smaller species such as *R. impeditum* and *R. racemosum*, and also for the evergreen azalea section. Half-ripe sideshoots are taken in June or July and inserted in pots or pans, using a compost of peat and lime-free sand. The receptacles are placed in a propagating frame.

RHUS *R. typhina* (sumach) is increased by root cuttings about 3 cm (1½ in) long which are secured in winter and inserted singly in small pots. Rooted suckers may be transplanted in autumn.

RIBES The flowering currants are easily propagated from hardwood cuttings about 15 cm (6 in) long inserted in the open in autumn.

ROBINIA Seed stored dry is sown in the open in spring. Root cuttings about 8 cm (3 in) long will grow in the open when planted in spring. Suckers are usually freely produced and only require transplanting. Grafting is the means by which varieties of *R. pseudoacacia* are propagated.

ROMNEYA (Californian tree poppy) The species, *R. coulteri* and *R. trichocalyx*, are propagated from root cuttings which are cut into lengths of about 2.5 cm (1 in) and are inserted horizontally in pots or boxes of sandy compost in winter. Seed may be sown under glass in spring.

ROSMARINUS (rosemary) Readily increased from half-ripe cuttings inserted in a cold frame in August.

RUBUS The ornamental brambles are increased by division or by tip-layering. *R. cockburnianus* (syn. *R. giraldianus*) is increased by root cuttings 3 cm (1½ in) long, which are inserted singly in pots placed under glass.

RUSCUS (butcher's broom) Propagate by dividing the creeping rootstock in spring.

RUTA (rue) Cuttings of firm young growths can be rooted in a frame in summer.

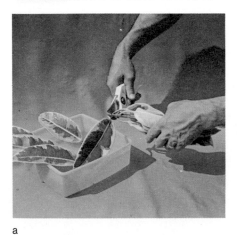

a

b

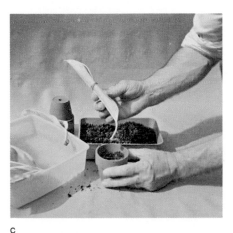

c

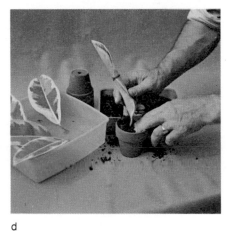

d

e

Plate 3. Leaf bud cuttings of *Ficus elastica* 'Tricolor'. Select a suitable stem and cut it up into 1 to 2.5 cm (½ to 1 in) sections (*a*). Each section should contain a leaf and its bud (*b*). After dipping the base of each cutting in hormone rooting powder insert it in a pot of cutting compost (an elastic band placed around the leaf will make handling easier) (*c*). Firm the compost into place (*d*) and hold the cutting upright by inserting a small length of split cane through the leaf tunnel. Root in a propagator.

a

b

c

d

e

f

Plate 4. Air layering *Ficus elastica*. This method is useful for reducing the size of plants which have outgrown their situation. Fasten the top cluster of leaves together with an elastic band to make the job easier (*a*). Remove as many leaves as is necessary to produce 10 cm (4 in) of clear stem (*b*). Make an upward incision at an angle of 45°(*c*). Fill the incision with moist sphagnum moss (*d*). Pad more moss around the stem and loosely hold this into place with cotton (*e*). Cover the pad with polythene held in place with adhesive tape (*f*) (*g*) and (*h*).

g

h

SALIX (willow) One of the easiest plants to increase from almost any type of stem cutting.

SALVIA (sage) The shrubby species are easily increased from softwood cuttings in a propagating frame.

SAMBUCUS (elder) The ordinary species are easily increased from seed which is stored dry and sown in the open in spring. Special varieties are raised from hardwood cuttings with a heel, taking care not to expose the pithy centre. These are inserted in the open. Softwood sideshoots will also strike in a propagating frame in July.

SANTOLINA (lavender cotton) Softwood sideshoots about 8 cm (3 in) long root readily in a propagator in summer.

SARCOCOCCA Easily increased by division, semi-ripe cuttings or by seed sown in spring.

SCHIZOPHRAGMA (climbing hydrangea) Usually increased from cuttings with a heel rooted in a propagator in summer.

SENECIO The shrubby species such as S. greyi and S. laxifolius and the hybrid S. 'Sunshine' root from cuttings of half-ripe sideshoots inserted in a propagator in August. Hardwood sideshoots can be rooted in a cold frame in autumn.

SEQUOIA (Californian redwood) Seeds can be sown in a frame in spring; cuttings of firm young growths can be rooted in a propagator or under mist in summer.

SEQUOIADENDRON (wellingtonia, giant redwood) Cuttings of firm young growths can be rooted in a propagator or under mist in summer. Seeds can be stratified for eight weeks and sown in a cool greenhouse.

SKIMMIA Ripened sideshoots taken in autumn strike in a cold frame but are rather slow. Cuttings of firm young growths inserted in a propagator in summer will root more quickly. Layering in autumn is another method.

SOLANUM The shrubby climbing species such as S. jasminoides are readily increased from cuttings rooted in a propagator in early summer.

SOPHORA Usually increased from seed sown in spring in gentle heat; summer cuttings rooted in a propagator and grafting on seedling stocks for S. japonica 'Pendula'. Suitable stems may be layered in spring.

SORBUS Seed sown outdoors in spring after being stratified during winter provides a means of increase but hybrids are usually grafted on to seedlings of the rowan (S. aucuparia).

SPARTIUM S. junceum (Spanish broom) is easily increased from seed sown in spring under glass.

SPIRAEA All the spiraeas can be propagated from cuttings either of the hardwood type taken in the autumn about 20 cm (8 in) long with or without a heel and planted in the open, or softwood cuttings struck in a propagating frame. Division is used for the dwarf stooling kinds such as S. douglasii and S. japonica and its varieties.

STACHYURUS Propagated by cuttings taken with a heel in July and inserted in a propagating frame; or by seed sown in spring or autumn.

STAPHYLEA (bladder nut) Increased by seed sown in cold frame in spring or autumn, by cuttings taken in autumn and inserted in a cold frame, and by suckers removed and planted while dormant.

STEPHANANDRA Propagation is by division or summer cuttings.

STEWARTIA Increase is by seed which should be soaked for a few days before sowing in a cold frame. Cuttings with a heel will strike in heat if taken in the autumn. Suitable stems may be layered in autumn.

STRANVAESIA Half-ripe cuttings will strike in gentle heat. Seed may be sown outdoors in autumn or stratified and sown in spring.

STYRAX Usually raised from seed, but sumer cuttings will strike in a propagating frame.

SYMPHORICARPOS (snowberry) Easily increased by hardwood cuttings taken in October and planted in the open, or by division.

SYRINGA (lilac) Tongued young shoots are layered in spring. Varieties are also grafted on seedlings of S. vulgaris (common lilac). This is carried out under glass in spring. Privet, which is sometimes used as a stock, should be avoided for it suckers badly. Some varieties can be rooted from half-ripe cuttings inserted in a cold frame in July.

TAMARIX Hardwood cuttings 20 to 25 cm (8 to 10 in) long are planted in the open in sheltered positions in autumn.

TAXODIUM (swamp cypress) Cuttings of firm shoots can be rooted in a frame in autumn; seeds can be stratified for eight weeks and then sown in a frame.

TAXUS (yew) Stratified seed is sown in the open in spring. Softwood terminal shoots inserted in pots may be rooted in a propagator or under mist.

TEUCRIUM (germander) The shrubby species root from small soft sideshoots taken in June or July and planted in a propagating frame.

THUJA AND THUJOPSIS Seed stored dry over winter is sown in spring in the open. Cuttings of sideshoots with a heel can be rooted in a cold frame if planted in September.

TILIA (lime, linden) The species are raised from seed where available, stratified and sown outside in spring or in pots or trays under glass. Layering is used for both species and varieties. Stock plants for this purpose are cut down and the young shoots layered annually in spring after bending.

TRACHELOSPERMUM (Chinese jasmine) Cuttings of young shoots strike in a propagator in spring or summer.

TSUGA (hemlock) Usually increased by seed sown outdoors in spring. Summer cuttings will also strike in a cold frame. Grafting of certain varieties on seedling stocks is also practised.

ULEX (gorse) Stratified seed sown outside in spring or in pots placed in a cold frame germinates freely. Sideshoots taken in July are inserted in a propagating frame and kept syringed frequently. Alternatively root under mist.

ULMUS (elm) The species are raised from seed sown in the open immediately it is ripe. Varieties are grafted on to seedlings of *U. glabra* (witch elm) in the spring.

VACCINIUM Seed of the different species is sown on sifted peat under glass in spring. Layering in the autumn is a quicker method, each young shoot being twisted before being pegged down. Cuttings of semi-mature sideshoots are inserted in a propagator in August. Divide established plants in spring or autumn.

VIBURNUM Several species are increased from seed sown in pots in spring and placed in a cold frame or greenhouse. Layering is another method used for *V. opulus* (guelder rose) and its varieties. Half-ripe shoots of certain species such as *V.* × *burkwoodii*, *V. carlesii*, *V. farreri (fragrans)* and *V. tinus* (laurustinus) strike readily in a propagator in July or August.

VINCA (periwinkle) Division in early spring is a simple method. Soft tips root easily in a propagator in summer and cuttings of ripe shoots can be rooted in a cold frame in the autumn.

VISCUM (mistletoe) Notoriously difficult to establish. Very ripe berries can be pushed into nicks cut on the underside of host tree branches in spring. Host trees include: apple, hawthorn, oak, larch, ash.

VITIS (vine) Varieties of the grape vine and *V. coignetiae* can be increased from hardwood cuttings 15 to 20 cm (6 to 8 in) long inserted in a cold frame in October.

WEIGELA Cuttings of firm young shoots can be rooted in a propagator in summer or in a cold frame in autumn; suitably pendent stems can be layered in autumn.

WISTERIA *W. sinensis* can be raised from seed sown in heat in early spring. Varieties are root-grafted on to these seedlings when established in winter. Young shoots may be layered into pots in autumn.

YUCCA Root cuttings 5 to 8 cm (2 to 3 in) long are secured in winter and inserted in boxes of sandy soil which are placed in heat. Rooted suckers may be removed in spring.

ZENOBIA Dried seed is sown in pans filled with sifted peat in spring. Layering is another method and is carried out in autumn.

15
How Roses are Increased

This valuable group of plants can be propagated by several methods, such as budding, by cuttings or by layering. When roses are budded this must be done on rootstocks specially grown for the purpose.

Rose stocks

ROSA CANINA (wild or dog rose) Although this stock is very widely used and its hardiness, strong root system and ease of transplanting make it a popular choice, its performance is somewhat inconsistent and poor results are not uncommon. This is probably due to the fact that *R. canina* is usually raised from seed which causes variability. Better results have been achieved with certain strains of this species such as Brogg's, Heinsohn's Record and Schmidt's Ideal. With strains like these, *R. canina* produces vigorous yet sturdy growth and roses worked on it have great lasting qualities. *R. canina* is also widely used for standard roses.

R. canina is easily propagated from cuttings of firm young shoots 23 to 25 cm (9 to 10 in) long. All the buds are removed except a few at the top. The cuttings may be prepared during the winter, tied in bundles and laid in sand for spring planting.

The rooted cuttings are usually lifted the following autumn and have their shoots cut back and rootlets trimmed off the stem above the main root. They are then replanted for budding the following summer.

It should be noted that the root system of cutting-raised *R. canina* stocks is not so sturdy or as deep as that produced by plants raised from seed. The plants are reputedly shorter-lived as well.

For standard roses stocks of *R. canina* are secured by collecting plants with tall straight stems from the hedgerows. Strains such as Pfander are often preferred by commercial growers.

RUGOSA This is a vigorous stock but roses on it do not live so long as when they are budded on briar. It is used mainly for standards, succeeds on light soil and, owing to its mass of fibrous roots, transplants readily. Propagation is very easily effected from cuttings, which are treated in the same manner as those of the wild rose. Rugosa stocks for standards are not as a rule grown in this country, but are imported from Holland (as are most seedling Canina stocks).

LAXA A stock which is recommended for light soils and also those of a calcareous nature. Owing to its thin bark it is easy to bud and is ready for budding early in the season. Laxa is used for bush roses but it does produce a lot of suckers. This stock is propagated from cuttings or suckers.

MULTIFLORA Valuable as a rootstock on light, dry soils. The bark is thin and the shoot arising from the bud may have to be supported in the early stages of growth to prevent it from falling out. Multiflora stocks may be raised from seed or cuttings, though the latter may be difficult to transplant successfully in dry conditions.

The stocks mentioned above may be propagated as described or obtained from nurserymen. For budding, selected strong plants are usually planted in rows about 75 cm (30 in) apart and 30 cm (12 in) allowed between the stocks.

Budding

This operation is carried out in summer from June onwards when the bark lifts easily. Timing, however, has considerable influence on the quality of the maiden growth. Thus, roses budded in late August or September are usually inferior in their first year to those budded earlier in the season. For bush roses the buds are inserted right at the base of the stem slightly below soil level, the soil being scooped away from the plants for this purpose. It is claimed that the buds secured from stems which are carrying faded blooms are the most satisfactory, but at any rate the buds near the centre of the shoots are preferable. If several varieties are being budded, each lot of stems from the same variety should be tied together, labelled, and stood in water until required.

The ordinary shield method of budding is used, and the buds secured with special rubber patches or plastic tape. At one time raffia was very popular. Nine out of ten of the buds should grow and this is indicated after a few weeks. Those buds which then appear green and plump have taken, but failures soon show signs of shrivelling and turn brown.

The following February the stocks are cut back close to the buds. Subsequently, the young shoots should be kept staked as they grow, as they are easily torn off by the wind. Remove suckers during the growing season and keep the land free from weeds. Roses propagated in this way may be planted in their permanent positions in the autumn and should then be cut back within 5 or 8 cm (2 or 3 in) of ground level.

Stocks for budding as standards are usually planted in the late autumn. These vary in height; thus for weeping standards they should be 2 to 2.5 m (6 to 8 ft) tall, ordinary standards are 1.25 to 1.4 m (4 to 4½ ft) and half-standards about 1 m (3 ft). As a rule all top growth is cut away and two shoots only, opposite each other, are allowed to grow at the stem apex. In the summer a bud is grafted on each of these laterals as close to the main stem as possible. These buds are inserted without making the usual cross-incision, that is, the cut that forms the top of the T. The omission of this cut reduces the risk of the lateral being broken off and, although it is a little more difficult, the bud can be worked sideways.

The method described above applies to the briar stock, but with the rugosa standard two or three buds are inserted directly on to the main stem just below the top growths. Standards require careful staking and the stake should extend above the point of union to give support to the head. Climbing and rambling roses are budded at ground level in the same manner as bush roses.

Cuttings

This is the simplest method for the amateur and is often the only one attempted. It is very suitable for most climbing and rambling roses as well as the majority of the old-fashioned roses.

There is considerable controversy regarding the merits of cuttings for increasing the popular hybrid teas and floribundas (or large-flowered and cluster-flowered roses). As a general rule most of these do best when budded on to a good stock. Plants from cuttings are often rather slow in coming into flower, and some varieties never develop into strong bushes when raised in this way. It is said that the red, pink and light-coloured varieties succeed better from cuttings than do those of yellow or fancy colours. Undoubtedly varieties do differ markedly in this respect and the keen gardener has an opportunity of experimenting with this simple method of propagation. One definite general advantage of cuttings is that there is no trouble with suckers arising from the rootstock.

Rose cuttings of most classes and varieties are not difficult to root if the right type of stem can be secured. Firm, well-ripened shoots are necessary, avoiding any with a large pith. As a rule, they are taken and planted in the autumn, but better results have been claimed when the work is done in late summer. The cuttings are made about 23 cm (9 in) long and are inserted outdoors 15 cm (6 in) deep and 15 to 20 cm (6 to 8 in) apart. It is an advantage to put a little sand at the bottom of the trench, and the cuttings should be trodden in firmly. Cloches placed over the cuttings provide protection against severe frost and are believed to promote rooting. In cold districts the cuttings may be inserted in a cold frame with advantage.

Usually the cuttings have produced roots by the following autumn and may then be transplanted.

Another method is to use softwood cuttings, which usually root freely. These consist of leafy shoot tips a few inches long which are cut off below a leaf. Remove the lower leaves and insert the cuttings in moist, sandy compost contained in a propagating frame and kept shaded. These cuttings may be rooted in a cold frame, but slight bottom heat gives quicker rooting and mist propagation can also be used. Rooted plants are potted and then gradually hardened off for planting outside.

Layering

This is an alternative method to cuttings and may be undertaken from July to September. A slit should be made in the part of the stem to be layered and this is then bent down to the ground and firmly embedded in the soil. By the autumn the layers have usually taken root and may then be severed and transplanted. Sometimes insufficient roots have been produced and it is necessary to leave the layers for another year.

Rambling roses may be tip-layered in the same manner as blackberries. The tip of the young shoot is usually buried in the soil to a depth of 10 cm (4 in). A new shoot grows up from the soil and is transplanted when well rooted.

Air layering is also recorded as being a successful method of propagating roses. It may be particularly valuable in increasing large old bushes.

Suckers

Roses growing on their own roots sometimes produce suckers which are quite suitable for propagation. These are simply lifted and transplanted during the dormant season.

Seed

The various ornamental rose species such as *R. hugonis* and *R. moyesii* may be propagated from seed, although there is always some risk of cross-pollination resulting in hybrids unless each species is well isolated. The hips should be collected in the autumn when ripe, cut open, and the seed washed out. Stratify the seed in sand until the spring, when it is sown in boxes or pans placed in a cold frame or greenhouse. The species may also be propagated fairly readily from cuttings and, of course, any of the other methods used for the general run of roses such as suckers, layering and budding.

16
Propagation of Herbaceous Perennials and Water Plants

Hardy herbaceous perennials

Stem cuttings are the most important method of increase. Usually these consist of cuttings made from the young basal shoots in spring, rather like chrysanthemum cuttings from stools. Sometimes for propagating purposes roots of certain herbaceous perennials such as lupins and delphiniums are brought into greenhouses to forward the growth of the basal shoots. Such cuttings are usually struck in March and April in cold frames. Other methods of increase are division, root cuttings and seed.

How the different kinds are increased

ACANTHUS (bear's breech) Root cuttings are the usual method, but division in spring is also successful. Seed is another means of increase.

ACHILLEA (yarrow) Easily increased by division in spring or autumn. Soft cuttings strike readily in spring. The species come true from seed.

ACONITUM (monkshood) Sow seed outside in April or May. Divide in spring or autumn.

ADENOPHORA (ladybell) Increased by careful division in spring. Also by seed sown in autumn or spring in pans placed in a cold frame.

ADONIS (pheasant's eye) Careful division in early spring is effective. Seed sown when ripe is a second method.

AGAPANTHUS (African lily) Divide established clumps in spring. Seeds may be sown but will take three years to flower.

ALCHEMILLA (lady's mantle) Divide established clumps in autumn or spring. Seeds can be sown outdoors in spring.

ALSTROEMERIA (Peruvian lily) Divide the roots in spring or sow seed thinly under glass in the autumn or early spring. Seedlings must be handled with great care and disturbed as little as possible.

ANCHUSA (bugloss) Easily increased from root cuttings about 10 cm (4 in) long planted in the open or in a cold frame in spring.

ANEMONE (windflower) The popular *A. hupehensis* (syn. *A. japonica*) and its varieties are usually increased from root cuttings in spring. Division in the autumn is another method, but the rooted pieces should preferably be potted.

ANTHEMIS (chamomile) Readily propagated from soft cuttings in spring or by division in spring or autumn.

ANTHERICUM (St. Bernard's lily) Sow seed in a cold frame as soon as it is ripe. Divide the plants in spring.

AQUILEGIA (columbine) The species can be increased from seed if plants are kept well isolated from one another. Excellent strains in mixed colours can also be raised from seed. Sow in the open in spring. Division of named varieties may be carried out in spring.

ARISARUM Divide the roots in spring.

ARTEMISIA (wormwood) Mainly increased by division, though cuttings can be rooted in a propagator in summer.

ARUM Increase by division in autumn.

ASCLEPIAS (milkweed) Propagated by division in spring or by seed in moderate heat.

ASPHODELINE Root division should be carried out in autumn or spring.

ASTER (Michaelmas daisy) All varieties are easily increased by division or by spring cuttings.

ASTILBE (goat's beard) Divide during the dormant period or sow seeds under glass in February.

ASTRANTIA (masterwort) Easily increased by root division in spring.

BAPTISIA Increased by seed or division.

BERGENIA (elephant's ears) Divide in spring or autumn.

BRUNNERA Divide mature clumps in spring or autumn.

BUPHTHALMUM (yellow oxeye) Propagate by division in autumn or by seed sown outdoors in spring.

CAMPANULA (bellflower) The tall herbaceous types are easily increased by division or spring cuttings. Seed may also be used.

CARNATION Border carnations are increased by layering. See Chapter 13.

CATANANCHE (Cupid's dart) Increase by division or seed in spring.

CENTAUREA (cornflower) Easily increased by seed or division.

CENTRANTHUS (valerian) Perennial species are increased by seed sown in slight heat in spring, or by division in autumn.

CEPHALARIA (giant scabious) Propagated by seed sown outdoors in spring.

CHELONE (turtle-head) Increase by autumn division, seed sown in spring or by cuttings rooted in summer in a cold frame.

CHRYSANTHEMUM *C. maximum* is readily increased by division or by spring cuttings.

CIMICIFUGA (bugbane) Divide in spring or sow seeds when ripe.

CODONOPSIS (bell-wort) Easily raised from seed, or by cuttings in late summer.

CONVALLARIA (lily-of-the-valley) Divide mature clumps in autumn. Seeds (when available) can be sown outdoors in spring.

COREOPSIS (tickseed) The perennial species are raised from seeds sown in the open in spring; or by division in spring or autumn.

CORTADERIA (pampas grass) Divide mature clumps in spring; sow seeds in a temperature of 16°C (60°F) in spring.

CORYDALIS (fumitory) Perennial species can be increased by seed or division.

CREPIS (hawk's beard) Divide mature clumps in spring; sow seeds in a frame in spring.

CROCOSMIA (montbretia) Easily increased by division in autumn or spring.

DELPHINIUM Propagate from basal cuttings taken with a heel in a slightly heated greenhouse or cold frame. Seed sown in heat in January and planted out in April or May will produce plants flowering in the autumn. Seed may also be sown in spring or early summer in the open.

DICENTRA (bleeding heart) Divide or take cuttings in spring.

DICTAMNUS Divide in spring or autumn; take root cuttings and insert in a frame in spring; sow seeds outdoors in summer or early autumn.

DIERAMA (wandflower) Sow seeds in spring. Divide mature plants in autumn.

DIGITALIS (foxglove) The perennial species are increased by seed or division.

DORONICUM (leopard's bane) Easily increased by division in autumn.

ECHINOPS (globe thistle) Root cuttings planted in a cold frame in spring are effective; or divide in spring or autumn.

EPILOBIUM (willowherb) Readily increased by seed or division.

EPIMEDIUM (barrenwort) Divide clumps in autumn.

EREMURUS Propagate by root division in spring. Seed can be sown but is often slow to germinate and seedlings take three years to reach flowering age.

ERIGERON (flea-bane) Divide in spring or autumn or secure basal cuttings in spring.

ERIOPHYLLUM Seed or division are the methods of increase.

ERYNGIUM (sea holly) Take root cuttings in spring and divide in spring or autumn.

EUPATORIUM (hemp agrimony) Hardy species are increased by seed sown outdoors in spring, cuttings in summer and by division in autumn or spring.

EUPHORBIA (spurge) Divide established clumps in spring or autumn; sow seeds outdoors in spring.

GAILLARDIA (blanket flower) Named varieties do not come true from seed and should be raised by division in spring or by cuttings taken in August or September and protected over winter under glass.

GALEGA Sow seed in the open in spring or increase by division in spring or autumn.

GERANIUM (cranesbill) Divide the plants in spring or autumn and sow seed in spring.

GEUM (avens) The hybrids should be increased by spring division, but several good strains can be raised from seed sown in spring.

GLAUCIUM (horned poppy) Sow seeds outdoors in

spring or early summer and move plants to their final positions in early autumn.

GYPSOPHILA (chalk plant) *G. paniculata* is easily raised from seed sown in heat in February or outdoors in April. Varieties must be raised either from soft cuttings secured in spring from plants grown in heat, or by root-grafting on to seedlings. The varieties 'Bristol Fairy' and *flore plena* (double form) are increased in this manner.

HELENIUM (sneezeweed) Can be divided almost any time during the dormant period and is also easily increased from spring cuttings.

HELIANTHUS (sunflower) Increase is similar to Helenium.

HELIOPSIS Easily increased by division.

HELLEBORUS (Christmas rose) Divide in spring or autumn. Sow seed immediately it is ripe in a cold frame or in the open.

HEMEROCALLIS (day lily) The clumps may be divided in spring or autumn.

HEPATICA Divide mature plants after flowering in spring. Seeds can be sown in a frame in spring.

HEUCHERA (alum-root) Divide in spring or increase by seeds sown in spring. Seedlings, however, are often variable.

HOSTA (plantain lily) Divide large clumps in spring or autumn.

HIERACIUM (hawkweed) Readily raised from seed or division.

INCARVILLEA Increased by division or seed, which takes three years to reach flowering size.

INULA (fleabane) Seed is an effective method of increase as is division in spring and autumn.

IRIS (flag) Seed is an important means of increasing this large genus, and most species can be so raised. Usually seed is sown when ripe and, after exposure to winter cold, germinates readily in spring in mild heat. Division is also widely used and is the only means of increasing varieties. The bearded irises should be divided immediately after flowering.

KIRENGESHOMA Divide mature plants in spring; sow seeds in a frame in spring.

KNIPHOFIA (red-hot poker) Normally increased by division in spring. The species come true from seed sown in spring.

LAMIUM (dead nettle) Divide clumps in spring or autumn.

LIATRIS (Kansas gayfeather) Easily increased from offsets secured and transplanted in spring. Seed may also be sown in spring.

LIGULARIA Divide mature clumps in spring or autumn; sow seeds outdoors in spring or early summer.

LIRIOPE Divide established clumps in spring.

LUPINUS (lupin) Easily increased from seed sown in heat in early spring or in the open later in the season. Named varieties are best increased from heeled cuttings secured when available in spring and inserted in pots. Lupins are difficult to divide, although this may be attempted.

LYCHNIS (campion) Easily increased by division in spring or autumn. Seed sown outside in spring provides another method.

LYSICHITUM Divide mature plants in spring; sow seeds in a frame in spring and keep moist.

LYSIMACHIA (loosestrife) Propagate by division in spring or autumn.

LYTHRUM (purple loosestrife)) Propagation is by division.

MACLEAYA (plume poppy) Methods of increase are by suckers detached and planted in a moist shady position; cuttings taken from leaf axils root readily in a shaded frame in summer.

MALVA (mallow) Propagate by seed or cuttings.

MONARDA (bee balm) Divide in spring or use soft basal cuttings.

MORINA (whorl flower) Divide mature clumps in spring; sow seeds in a frame in spring or autumn.

MYOSOTIDIUM (Chatham Island forget-me-not) Sow seeds individually in small pots of seed compost in spring. Place in a frame and plant out as soon as the root system has filled the pot.

NEPETA (catmint) Readily increased by division or by soft cuttings taken in summer or autumn.

OENOTHERA (evening primrose) Insert cuttings of the perennial species in a frame before flowering. Divide in spring. Seed is another method.

ONOPORDON (Scotch thistle) Sow seeds outdoors in early summer. Transplant to flowering positions in autumn.

OTHONNOPSIS Cuttings can be rooted in a propagator in summer.

PAEONIA (peony) The species should be raised from seed sown when ripe in cold frame. Sometimes germination is slow. Divide in autumn, ensuring that each piece of root has an 'eye'.

PAPAVER (poppy) The well-known oriental poppies are easily increased from root cuttings about 10 cm (4 in) long secured and planted in the open in spring.

PHLOX *P. paniculata* can be raised from seed, but its many varieties are increased by basal cuttings or by division in spring or autumn.

PHYSALIS (Chinese lantern) Natural increase by underground stems occurs rapidly. Lift and divide when necessary.

PHYSOSTEGIA (false dragonhead) Sow seed in a cold frame in spring or divide roots in autumn or spring.

PHYTOLACCA (pokeberry) Divide large plants in spring; sow seeds outdoors in spring.

PLATYCODON (Chinese bellflower) Propagation is by seed or division.

POLEMONIUM (Jacob's ladder) Divide in spring or autumn. Seed may be sown in spring.

POLYGONATUM (Solomon's seal) Propagate by division in spring or autumn.

POTENTILLA (cinquefoil) Increase by seed sown in the open in spring or divide in spring or autumn.

PYRETHRUM The pyrethrum is really a chrysanthemum. It is increased by division in spring or immediately after flowering.

RUDBECKIA (cone flower) Divide in spring or autumn. Basal cuttings root readily in a cold frame in spring. Seed may be sown out of doors in spring.

SANGUINARIA (bloodroot) Divide mature clumps in autumn; sow seeds in a frame in spring in a mixture of peat, sand and leafmould.

SALVIA (sage) Division in spring or seed sown in spring are the usual methods.

SCABIOSA Divide in spring only. Seed is another method but varieties do not come true.

SIDALCEA Easily increased by division in spring or autumn.

SMILACINA Divide mature clumps in spring or autumn; sow seeds in a frame in spring.

SOLIDAGO (golden rod) Easily increased by division in spring or autumn.

STACHYS Increase by division in autumn or spring.

STOKESIA Usually propagated by seed or division.

SYMPHYTUM (comfrey) Divide established clumps in spring.

TELLIMA Divide mature plants in spring.

THALICTRUM (meadow rue) Divide in spring just as growth starts or sow seed in the open in spring.

THERMOPSIS Seed is the best method of increase. Plants may also be divided in spring.

TIARELLA Divide mature clumps in spring; seeds can be sown outdoors or in a frame in spring.

TRADESCANTIA (spiderwort) Basal cuttings root readily in spring or plants may be divided at the same time.

TRICYRTIS (toad lily) Sow seeds in a peaty compost in a cool greenhouse in spring; remove offsets and replant in spring or autumn.

TRILLIUM (American wood lily) Increased from seed sown when ripe. Mature clumps may be divided in autumn.

TROLLIUS (globe flower) Propagate by division just after flowering or sow seed when ripe.

UVULARIA (bellwort) Divide in spring or autumn; sow seeds in peaty compost in a frame in spring.

VERBASCUM (mullein) Readily increased from root cuttings in spring, or division, or by seed sown in spring.

VERONICA (speedwell) Divide in spring or autumn, or take basal cuttings in spring.

Half-hardy herbaceous perennials

CHRYSANTHEMUMS (*Chrysanthemum* × *morifolium* varieties) The florist's chrysanthemum is probably the most popular cut flower grown in this country, and for this purpose is widely cultivated by both amateurs and professional nurserymen. All-the-year-round production has resulted in the development of a new propagation technique, but as traditional procedure is still adopted by some growers both methods are described here.

Propagation by cuttings applies to both methods and the first necessity for success is to secure these only from healthy plants that are true to type and variety. This involves careful inspection during the growing season and the propagator should familiarize himself with the symptoms of the common pests and diseases of chrysanthemums such as eelworm, various types of virus, like aspermy, and fungus diseases such as verticillium wilt.

In the traditional method plants for propagation are carefully labelled after flowering and then cut down to within 15 cm (6 in) of soil level.

Plants for propagation are usually lifted about the end of October, and it is a good idea then to wash all soil completely off the roots to get rid of soil pests, particularly slugs and their eggs. All leafy growths on the stools should be trimmed off. The stools may be placed in boxes, and clean light soil, John Innes or a soilless potting compost packed around them. Pot plants may be either left in their pots or knocked out and laid in trays of compost.

During the winter sufficient artificial heat should be given to afford protection against frost. Unheated frames should be covered with mats or straw during severe frost. Only slight heat is

125

necessary to start the plants into growth and the young basal shoots are used as cuttings.

The most suitable time for taking cuttings varies with the type of chrysanthemum and also with the purpose for which the new plants are intended. Thus, cuttings of the large Japanese varieties which are often grown for exhibition are usually secured in December or January. Midseason and late decoratives which flower from October to December are, as a rule, taken in January or February, while propagation of the early-flowering types is normally carried out in February or early March. For dwarf pot plants, however, cuttings need not be secured until early May.

In unheated houses or frames growth may be delayed and one may have to wait until shoots for cuttings are produced. Some varieties throw up few or no basal growths so stem cuttings (as distinct from the basal cuttings arising from stools) must be used. Traditionally basal stool cuttings were supposed to be superior to stem cuttings but this has largely been disproved. In fact in the alternative method of cutting production as described below stem cuttings are used exclusively.

For this purpose rooted cuttings are planted out 12 cm (5 in) square usually in the greenhouse beds or borders. Subsequently as they grow they are stopped several times to induce the production of laterals. In this way they become stock plants for cutting production. Further experience of this method has indicated that it is better to grow stock plants with a 15 cm (6 in) stem, as cuttings borne nearer the ground are more liable to infection from pests and diseases. To achieve this the stock plant rooted cutting is simply allowed to grow 15 cm (6 in) tall before stopping.

The rate of cutting production from stock plants is related to temperature and to maintain an adequate rate of growth during the winter months a minimum night temperature of between 13 and 16°C (55 and 60°F) is recommended. However, it must be remembered that autumn- and winter-flowering chrysanthemums are naturally short-day plants (that is, they require short days to produce flower buds) and under short-day winter conditions when the temperature exceeds 13°C (55°F) the stock plants will tend to produce buds and flowers on short stems instead of vegetative growth.

Premature flower development in stock beds can be prevented by subjecting the beds to a period of artificial lighting during the night, so that dark periods never exceed 7 hours. This means that if cuttings are required between October and April the beds must be illuminated for periods of between 2 and 5 hours per night. During this season also cutting beds for rooting must be subjected to similar light periods to prevent bud initiation. Lighting for this purpose is easily effected by suspending ordinary 100 watt pearl bulbs above 1.25 m (4 ft) beds at a height of 1.4 m (4½ ft) and 2 m (6 ft) apart. Further information on this subject may be obtained from books on all-the-year-round chrysanthemum production.

The chrysanthemum cutting is one of the easiest to root and these notes apply to both basal cuttings from stools and stem cuttings from stock plants. Cuttings may be rooted in pots, trays or greenhouse benches. The ideal rooting compost should allow free drainage. The following mixture is recommended—3 parts by volume of coarse sand (3 mm (⅛ in) grist) and 2 parts granulated peat. Bottom heat in the beds or below the containers of about 18°C (65°F) is recommended with an aerial temperature of about 16°C (60°F). Heavy permanent shading should be avoided over cutting beds even in summer but temporary shading such as provided by white muslin may be necessary especially during the first few days after insertion if bright sunshine is experienced.

In the taking of cuttings a simplified procedure is now adopted. This consists of breaking off (not cutting off) the cuttings from the stock plants, taking care to leave at least three leaves at the base of each stem for further cutting production. Cuttings should be kept to a standard length of 5 to 6 cm (2 to 2½ in) in winter and for the rest of the year 4 to 5 cm (1½ to 2 in) is suitable. Without further preparation, except hormone treatment to promote more rapid rooting, the cuttings are inserted and well watered in. Subsequently, the cuttings are sprayed over at intervals to maintain humidity according to the time of year. Automatic misting may be used instead, but apart from saving labour has no other advantages.

The propagation of early flowering chrysanthemums may be carried out in a similar manner to autumn and winter varieties. Cuttings for the former are usually taken in March and for them artificial light is not necessary if the propagating temperature does not exceed 10°C (50°F).

When cuttings are rooted the subsequent treatment varies according to the proposed method of culture. Thus, early flowering chrysanthemums are often bedded out in a frame where they are protected from frost until they are planted in the open usually in April or May. Plants for growing and flowering in large pots are potted first in 8 cm (3 in) pots and later transferred to their final 20 to 25 cm (8 to 10 in) pots. All-the-year-round varieties are usually bedded out in greenhouses when rooted, or set in pans for sale as pot plants.

DAHLIAS Dahlias, like chrysanthemums, are usually propagated from cuttings, and healthy plants true to type should be selected for the production of suitable material. When the foliage shows severe frost damage it is time to lift the plants from outdoor beds, cut off the stems close to ground level, and thoroughly dry the tuberous roots. Each root should be carefully labelled and stored for the winter.

The essential storage conditions are dryness and freedom from frost, and any place which satisfies these requirements, such as a shed or under-stage area in the greenhouse, is suitable. The roots, however, should be surrounded with soil, sand, peat or sawdust in a fairly dry condition. Before the roots are stored they should be carefully examined and any decayed or damaged portions should be cut away and the cut surface sprinkled with sulphur.

Usually at about the New Year, the tuberous roots are arranged fairly close together on the staging of a greenhouse, or they may be placed in boxes. Moist peat or light, open soil should be packed around them. Artificial heat is necessary for early cuttings, but they will be produced later in a cold house. The house should be light enough to promote sturdy growth.

The new shoots on dahlias arise, not from the so-called tubers, but from the bases of the old stems. It is an advantage to allow the first growths produced to form four joints and then cut them back to the second node. This method induces further shoots to arise which soon become suitable for cuttings.

Sturdy growths are best for cuttings, and thin, lanky shoots should be avoided. Sever the cutting below a node or with a small piece of the old stem attached to form a heel. Remove the lower leaves and insert the cuttings in moist sandy compost.

As a general rule 9 cm (3½ in) pots are most suitable for rooting cuttings and about half a dozen cuttings should be inserted around the edges of each pot. The pots may be placed on the staging of a warm house or in an open propagating frame. Very humid conditions are likely to induce damping off, but bottom heat is beneficial.

Cuttings kept in a temperature of 13 to 16°C (55 to 60°F) root in two to three weeks, and in cooler conditions rooting is less rapid. Rooted plants are potted in 8 cm (3 in) pots using JIP₁ or a soilless potting compost. They are kept in a reasonably humid spot for a few days but are gradually exposed to cooler and more open conditions.

Dahlias are usually stood in a cold frame for a few weeks prior to planting outside. Very often they are set in the open in May from the 8 cm (3 in) pots. If conditions are unsuitable for planting, they should be potted on into larger pots and thus kept growing.

Increase by division Division is an effective method of propagating dahlias for the private gardener or commercial grower who has no glass. The dry tubers may be left until April, when they are divided up with the aid of a sharp knife. Make sure that one or more tubers are attached to a piece of stem with a bud. Six plants can often be secured from an average-sized root. The divided portions are then planted out sufficiently deep to keep them safe from frost damage.

Seed Seed offers a third method of increase and, although it is not suitable for the general run of named varieties, yet within recent years excellent strains, which come true to type and colour from seed, have been developed. These include the Coltness Gem hybrids and other dwarf bedders.

Where heated greenhouses are available, the seed may be sown in gentle heat in February or March. Sow thinly in boxes or pans filled with John Innes or a peat-based seed compost, and cover very lightly. When the seedlings are large enough to handle they are pricked out in boxes or potted up singly. Subsequent treatment is the same as for plants raised from cuttings.

A few other plants such as *Aster pappei, Cuphea ignea, Dimorphotheca* (several species) and *Felicia amelloides* may be classed as half-hardy peren-

127

nials. When rooted they may be potted and retained over winter in a heated greenhouse for planting outside in May.

Water plants

The term water plant as it is used here refers not only to some aquatics that actually grow in water, but also to those which require a moist situation, such as the margin of a pool or stream.

The majority of these plants are easily propagated by division. This is usually carried out in spring when the divided roots are set in their permanent positions. A number of plants are also increased by soft cuttings which root quickly in moist compost kept in a temperature of about 16°C (60°F). Seed is also used for several kinds and is usualy sown in spring. The usual method is to sow in shallow pans filled with sifted loam. Cover the seed lightly with sand and then stand the pans in a tank so that they are just covered with water. As soon as the first true leaves appear the seedlings are transplanted into pots or boxes. A temperature of about 16°C (60°F) is advantageous during the propagating period.

How important kinds are propagated

ACORUS (sweet flag) This plant grows in shallow water and is increased by division.

ALISMA (water plantain) Usually increased by division in spring or autumn but seed may also be used.

APONOGETON (water hawthorn) Increased by seed sown in spring or division about the same time.

ARUNCUS Propagated by seed or division.

BUTOMUS (flowering rush) Propagated by division.

CALLA (bog arum) Increased by division.

CALTHA (marsh marigold) Increased by division.

CYPERUS (umbrella grass) Propagate by division or seed sown in gentle heat.

DECODON Increase is by division or cuttings.

ERIOPHORUM (cotton grass) Seed and division are the methods of increase.

GLYCERIA (manna grass) Propagated by division.

GUNNERA (prickly rhubarb) Propagated by division.

HOTTONIA (water violet) Increased by division.

HOUTTUYNIA Usually divided in spring; seed is another method of increase.

HYPERICUM (marsh St. John's wort) Increase

H. elodes by division.

IRIS (flag) This genus includes many beautiful waterside plants propagated by seed or division.

JUNCUS (rush) Special varieties of the rush are grown and increased by division.

LOBELIA Species such as *L. fulgens* are propagated by division.

LYSIMACHIA (loosestrife) Divide or take cuttings in spring.

MARSILEA Increase is by division.

MENTHA (water mint) Easily increased by division.

MENYANTHES *M. trifoliata* (bog bean) is increased by division in spring.

MIMULUS (monkey flower) Easily raised from seed, division or cuttings.

MYOSOTIS (water forget-me-not) Raised from seed or by division in spring.

NYMPHAEA (waterlily) Most water lilies are increased by division in May. The divided portions, each having at least one eye, are planted in pots or boxes and are submerged in shallow water. Heat is advantageous but not essential. A few species can also be raised from seed, which is preferably germinated in a warm house with the pans just covered with water.

NUPHAR (yellow water lily) Propagation is similar to that of the water lily.

ORONTIUM (golden club) Increase is by division or seed.

PELTANDRA (arrow arum) Increased by division in spring.

PONTEDERIA The species *P. cordata* is raised from seed or by division.

POTAMOGETON (pond-weed) Natural increase is very rapid.

RANUNCULUS (buttercups) Increased by seed or division.

RODGERSIA Increase is by division or root cuttings.

SAGITTARIA (arrowhead) Propagated by division of the roots or runners.

SAURURUS (lizard's tail) Propagate by seed or division.

SCIRPUS (club rush) Division of the suckers is the method of increase.

THALIA Increase by division in spring.

TYPHA (reedmace, false bulrush, cat-o'-nine tails) Natural increase is usually too rapid. Plants divide or transplant easily.

17
Multiplication of Alpines

The principal methods of propagating alpines are by seed, cuttings or division; a limited number may be increased by leaf or root cuttings.

Seed is a quick and cheap method and many species such as *Erinus alpinus* and *Iberis sempervirens* may be sown in spring when they will germinate quickly in gentle heat. With other species, however, it is better to sow in the autumn or winter and subject them to a cold period in the open before bringing the containers inside to germinate in the spring. In some cases germination may take up to a year or even longer.

Cuttings are usually made from soft young shoots taken when available during the growing season. The majority will root then quite readily in a cold frame. Rooted cuttings and seedlings are usually potted into 8 cm (3 in) pots and kept moist and shaded until established. The pots are often plunged in sand or ashes to avoid frequent watering. When well rooted in the pots they may be planted in their permanent positions.

How the different kinds are increased

ACAENA Easily propagated by division in spring or autumn. Seed is also suitable.

ACANTHOLIMON (prickly thrift) Cuttings with a heel in summer will root in a frame or, preferably, inserted around the sides of a pot. Shoots may also be partly severed at a joint by pressing downwards and then layered by packing sandy compost around them.

ACHILLEA (yarrow) Soft cuttings taken from June to September are easy to strike, and plants may be divided in spring or autumn.

ADONIS (pheasant's eye) Rock garden species may be propagated from seed sown when ripe. Spring division is another method.

AETHIONEMA (stone cress) Summer cuttings are the usual means of increase. *A. armenum* comes true from seed.

AJUGA (bugle) Species suitable for the rock garden are easily increased by cuttings or division. Seed is sown in spring.

ALYSSUM The general method of propagation is by soft cuttings. Several species may be raised from seed sown in February, including *A. argenteum*, *A. montanum* and *A. saxatile*.

ANAGALLIS (pimpernel) Methods of increase are soft cuttings in May or June, seed sown in February and division in May, the rooted portions being potted.

ANAPHALIS (pearly everlasting) Seed may be sown outdoors in spring. Division may be carried out in spring or autumn.

ANDROSACE Offsets of the species *A. geraniifolia*, *A. primuloides* and *A. villosa* may be potted up in August. *A. lanuginosa* is increased from soft cuttings. *A. lactea* can be raised from seed sown in February.

ANTHEMIS (chamomile) Readily increased by seed or division.

ANTHYLLIS (kidney vetch) *A. montana* is increased by heel cuttings from June to August.

ANTIRRHINUM (snapdragon) Cuttings root readily during the summer. *A. asarina* may be increased from seed sown in January.

AQUILEGIA (columbine) Seed is produced freely and if home saved should be sown when ripe. If different species are being grown there is a risk of cross-pollination.

ARABIS (rock cress) Cuttings in summer is the normal method, but division in September or March is suitable for several kinds. *A. blepharophylla* may be raised by seed sown in July.

ARENARIA (sandwort) These are increased by summer cuttings or by division in autumn. *A. montana* can be raised from seed sown in April.

ARMERIA (thrift) Divide in spring or take cuttings in July or August.

ARNEBIA (prophet flower) The species *A. echioides* is propagated by seed. Also by cuttings taken with a heel in autumn. Bottom heat is an advantage. Root cuttings will also grow.

ASPERULA (woodruff) *A. hirta* and *A. nitida* are increased from seed sown in March or April. They can also be increased by division in spring. Softwood cuttings taken after flowering provide another method and are the best means of propagating *A. suberosa*.

ASTER Soft cuttings in summer or division in autumn are alternative methods.

ASTILBE (goat's beard) Easily increased by division.

ASTRAGALUS (milk vetch) The prostrate species suitable for the rock garden are usually increased from seed sown in autumn.

AUBRIETA (rock cress) *A. deltoidea* varieties are increased by cuttings in August and September, or by division in autumn. Seed is suitable for mixed colours and is sown in March.

AURICULA The alpine species are readily increased by seed or offsets.

BELLIS (daisy) *B. perennis* and its varieties are easily increased by division in July.

BELLIUM Divide in August and replant in the open.

CALAMINTHA (calamint) Seed may be sown in January or soft cuttings rooted in May or June.

CALCEOLARIA (slipper flower) The dwarf species are increased by seed sown in the autumn and by division in spring.

CAMPANULA (bellflower) Almost all rock campanulas may be increased from pre-flowering cuttings and many kinds by division in spring or autumn. Several species come true from seed which is sown in spring, examples are *C. barbata, C. betulifolia, C. mirabilis* and *C. rotundifolia*.

CARDAMINE (cuckoo flower) Those suitable for the rock garden are easily increased by division. *C. pratensis* can also be propagated from leaf cuttings.

CARLINA (carline thistle) Increase is by seed sown in spring.

CELMISIA Propagate by cuttings or by division in spring.

CENTAURIUM Increased by seed sown when ripe in August or by summer cuttings.

CERASTIUM (snow-in-summer) Cuttings root readily in summer—or divide the plants in September.

CHEIRANTHUS (wallflower) The rock garden varieties are usually increased from cuttings in summer.

CODONOPSIS (bellwort) Easily increased from cuttings and by seed in spring.

CONVOLVULUS Increased by division in April and by summer cuttings. The species *C. cantabricus* can be raised from seed sown in April.

CORONILLA (crown vetch) Propagation is by summer cuttings and by layers.

CORTUSA The species *C. matthioli* is increased by seed sown in March or by division in the same month.

COTULA Divide in autumn and pot up the rooted pieces—or they may be planted on the rock garden.

COTYLEDON Can be raised easily from stem or leaf cuttings. Seed is likely to give variation.

CYANANTHUS Cuttings strike best in early summer well in advance of the flowering period.

CYCLAMEN Increase the rock garden species by seed and division of the clumps in spring.

CYPRIPEDIUM (lady's slipper) The usual method is by division in spring or autumn. Seed may be sown in February or March if available.

DIANTHUS (pink) The majority of these may be increased by summer cuttings. Several species such as *D. gratianopolitanus (caesius)* are readily increased by autumn division. The species *D. arenarius, D. deltoides, D. superbus, D. sylvestris* and *D. zonatus* can be raised from seed sown in February.

DRABA (whitlow grass) Offsets may be potted up in spring. Several species are increased from summer cuttings. *D. aizoon* can be raised from seed sown when ripe or in spring.

DRACOCEPHALUM Rock garden species are multiplied by division in spring and cuttings in May.

DRYAS (mountain avens) Easily increased by heel cuttings taken in August or by division of the naturally layered shoots.

ECHEVERIA Offsets can be removed in spring; seeds can be sown in a temperature of 16°C (60°F) in spring; individual leaves can be rooted in sandy compost in summer.

EDRAIANTHUS (American cowslip) Methods of increase are division in April or August or by seed sown in February. Plants from seed are usually variable.

EPILOBIUM (willow herb) Some species arise freely from self-sown seed. Collected seed should be sown in spring. Summer cuttings provide another method.

ERIGERON (fleabane) Several species are increased from seed sown in March. Summer cuttings root readily, and increase is possible by division in autumn.

ERINACEA (hedgehog broom) Take cuttings after flowering and insert around the edges of a pot. Seed sown in the New Year germinates best when exposed to frost.

ERINUS Easily raised from seed sown in spring and by division at any time.

ERIOGONUM Heeled cuttings taken from July to September are used to increase such species as *E. umbellatum* and *E. subalpinum*. Division is another method and seed comes true in most cases. It should be sown in spring.

ERIOPHYLLUM Increase by seed sown in autumn and by division carried out while the plant is in active growth.

ERODIUM (heron's bill) Easily propagated from summer cuttings.

FRANKENIA (sea heath) Normally increased by division in September or April.

GENTIANA A great many of the gentians may be increased from cuttings including *G. farreri*, *G. lawrencei*, *G.* × *macaulayi* and *G. veitchiorum*. Seed is also generally used. It should be sown when ripe and exposed to frost. Subsequently it usually germinates readily in mild heat. Several species may be divided after flowering. These include *G. acaulis*, *G. alpina*, *G. angustifolia* and *G. sino-ornata*.

GERANIUM (cranesbill) Seed sown in February is the general method of increase and applies to such species as *G. argenteum*, *G. ibericum* and *G. sanguineum* and its variety *lancastriense*.

GEUM (avens) Divide in March or September and replant in the open.

GLOBULARIA (globe daisy) Propagation is by spring-sown seed or division in autumn.

GYPSOPHILA (chalk plant) Easily increased by cuttings secured before flowering or by division in spring or autumn.

HABERLEA The methods of increase are seed and division. Leaf cuttings will also root taken as for Ramonda.

HELIANTHEMUM (rock rose) Summer cuttings secured from June onwards strike redily. *H. guttatum*, which is an annual, is raised from seed.

HELICHRYSUM (everlasting flower) Increase by seed sown in February. *H. bellidioides* and *H. plicatum* can be propagated from summer cuttings.

HIERACIUM (mouse ear) Sow seed in autumn or increase by division in spring.

HOUSTONIA (bluets) Readily increased by division after flowering in August or in spring. Seed is also used and should be sown in March.

HUTCHINSIA The well-known species, *H. alpina*, is increased by division in March or April by seed sown as soon as ripe.

HYPERICUM (rose of Sharon) Most species may be increased from summer cuttings. *H. coris*, *H. fragile* and *H. rhodopeum* can be raised from seed sown in February.

IBERIS (candytuft) Summer cuttings root easily, and seed or division are alternative methods.

IRIS (flag) A great many of the rock garden irises may be increased from seed, which is usually freely produced. Varieties, however, must be propagated vegetatively, which is done by division. As a rule the best time to divide the rhizomatous types is just after flowering.

JEFFERSONIA (twin-leaf) Sow seed when ripe or propagate by division.

LEONTOPODIUM (edelweiss) The species *L. alpinum* is increased by seed sown in spring or by division, again in the spring.

LEWISIA Division can be carried out with some species and should be attempted in spring. Seed sown when ripe is usually successful, but may give rise to variable seedlings.

LINARIA (toadflax) Many species are easily raised from seed sown on the rock garden. Summer cuttings or division in March provide other methods.

LINUM (flax) Take soft cuttings in June or July with a heel *L. flavum*, *L. monogynum* and *L. narbonense* may be raised from seed sown in spring.

LITHOSPERMUM Cuttings secured in July may be rooted in a mixture of peat and sand.

LYCHNIS (campion) Seed is usually produced freely and should be sown in February. Some species can be divided.

MAZUS Divide in spring or autumn or sow seeds in spring.

MECONOPSIS Sow seed in a peaty mixture as soon as ripe. Grow the seedlings in a shady position.

MENTHA (mint) This is easily increased by division in March.

MERTENSIA Readily increased by division in autumn or by sowing seed when ripe.

MIMULUS (musk) Cuttings taken in summer root readily in moist soil. Division may be carried out in spring. *M. cardinalis* and *M. lewisii* are examples of species which may be raised from seed sown in spring.

MORISIA Root cuttings about 2.5 cm (1 in) long are secured from June until August and are laid in moist sand. Seed is another method.

MYOSOTIS (forget-me-not) Normally raised from seed sown in April or by division in spring.

NIEREMBERGIA (cup flower) *N. repens (rivularis)* is propagated by division in August. *N. caerulea* and *N. frutescens* should be cut back after flowering; the young shoots produced are suitable for cuttings.

OENOTHERA (evening primrose) Heel cuttings taken in June or August usually strike easily. Several species may be increased by division. Some species such as *O. tetragona* can be raised from seed sown in early spring.

OMPHALODES (rock forget-me-not) Increased by division in spring or autumn; seed is also used.

ONOSMA (golden drop) Take cuttings after flowering in July. Seed should be sown in February.

ORIGANUM (marjoram) Easily increased by seed, cuttings or division.

OURISIA Seed and division in spring are the usual methods.

OXALIS (Cape shamrock) Seed will germinate in gentle heat in spring. Roots may be divided when planting or potting.

PENSTEMON (beard tongue) *P. confertus*, *P. glaber*, *P. hirsutus* and *P. humilis* may be increased by division in March. Soft cuttings are generally used. Certain species such as *P. eatonii*, *P. glaber* and *P. erianthera* may be raised from seed, which is usually sown in March.

PHLOX Secure cuttings after flowering in summer. *P. stolonifera* can be readily divided in spring or autumn. *P. subulata* can also be increased by division which is facilitated by sifting some fine soil among the growths a few weeks before the operation is attempted.

PHYTEUMA (horned rampion) Easily propagated by seed sown in spring or by division in autumn. The seed of *P. comosum* should be sown immediately it is ripe in September.

PLATYCODON The species *P. grandiflorum* may be divided in spring. It is also readily increased by sowing seed in March.

POLEMONIUM (Jacob's ladder) Propagated by division in autumn or by seed sown in spring.

POLYGALA (milkwort) Cuttings of young shoots will strike in summer in a frame.

POLYGONUM (knotweed) Increased by summer cuttings or by spring division.

POTENTILLA (cinquefoil) Methods of increase are sowing seed in spring, heel cuttings in summer for the shrubby species and species such as *P. alba*, and *P. cuneata*. *P. verna* may be divided in spring.

PRIMULA A great many primulas are easily raised from seed, which is best sown as soon as ripe. Species raised in this way include the bog primulas such as *P. japonica* and *P. pulverulenta* and various other species like *P. chionantha*, *P. cortusoides* and *P. frondosa*. Most primulas may also be propagated by division in autumn.

PULMONARIA (lungwort) Easily increased by division in spring or autumn.

PULSATILLA (pasque flower) Sow seeds in a frame in spring and transplant seedlings with as little root disturbance as possible.

RAMONDA Seed may be sown in peaty soil in September. Leaf cuttings are a popular method but each leaf must have a bud at its base, and is inserted to a depth of 2.5 cm (1 in) or so in sandy compost. Rooted offsets can often be pulled off in March.

RANUNCULUS Seed should be sown when ripe in May or June. The seed of some species such as *R. glacialis* and *R. geraniifolius* germinates best after exposure to frost when sown.

RAOULIA Division should be carried out carefully in late summer or autumn.

ROSA (wild rose, briar) The dwarf alpine roses such as *R. pendulina* are easily increased from summer cuttings.

ROSCOEA These are usually propagated by division in spring.

SAGINA (pearlwort) Increase is by seed or division in spring.

SAPONARIA (soapwort) Easily increased by soft cuttings in summer.

SAXIFRAGA (rockfoil) Most of the plants in this very large genus are not at all difficult to propa-

Plate 5. Layering *Magnolia stellata*. Select a suitable
pendent stem and cut it with a sharp knife to produce a
'tongue' (*a*). Peg the stem to the ground so that the
tongue is held open against the soil (*b*). Tie the shoot tip
to a cane pushed into the soil (*c*). Cover the layered
portion of the stem with a mixture of soil and peat (*d*).
The rest of the soil around the plant can be mulched at
the same time (*e*). The young layered shoot (*f*) should
grow away well the following season and can be removed
a year later.

a

b

c

d

Plate 6. Layering a rhododendron. Choose a flexible
shoot that can be layered without risk of breakage. Make
an incision in the stem (*a*) then twist and peg down the
shoot into a hollow made in the soil (*b*). Support the
shoot tip by tying it to a cane (*c*) and then topdress with
peaty compost (*d*).

gate. Division is the most important method and applies in particular to the encrusted and mossy types and others which produce rosettes or offsets. *S. oppositifolia* type is readily increased from cuttings in summer, and the 'Kabschia' section represented by *S. burserana* is best increased by very small cuttings secured after flowering.

SCUTELLARIA (skull cap) Easily increased by summer cuttings. *S. alpina* can also be raised from seed sown in September.

SEDUM (stonecrop) The sedums are among the easiest plants to propagate and many can be increased by simply scattering their broken stems and leaves over the soil; examples are *S. acre*, *S. kamtschaticum* and *S. middendorffianum*. Ordinary cuttings in summer provide the means in other cases. Some can be divided and seed may be used to increase a number of species such as *S. caeruleum* and *S. pilosum*. It is sown in February.

SEMPERVIVUM (houseleek) Mainly increased from offsets which are planted in the rock garden in May, but if secured in September they should be potted for spring planting.

SHORTIA Seed sown when ripe if available, or the plants increased by division.

SILENE (catchfly) Several species such as *S. maritima* and *S. schafta* are increased by seed sown in spring. *S. acaulis* and many others may be propagated from summer cuttings or by division in July or August.

SISYRINCHIUM (satin flower) Propagated by seed sown in February.

SOLDANELLA (moonwort) Sow seed immediately it is ripe. Plants may be divided in June.

SPIRAEA (meadowsweet) Easily increased by division which is best carried out in spring. Roots may be cut in pieces with a knife, taking care to leave at least one bud on each.

TEUCRIUM (germander) Summer cuttings are not difficult to root. *T. marum* and similar species should be taken with a heel. Division is used to propagate some kinds.

THYMUS (thyme) Readily propagated by soft cuttings in summer. Plants may be divided in spring or autumn.

TRIFOLIUM (clover) Normally raised from seed sown in August. Several species can also be increased by division.

TUNICA Cuttings taken in July or August root readily.

VERBENA (vervain) The usual method is by cuttings taken in late summer and autumn. When rooted these are potted and wintered indoors.

VERONICA (speedwell) The veronicas are quite easy from cuttings taken in summer. Many species can be divided in spring or autumn.

VIOLA (heartsease) Several species such as *V. biflora*, *V. blanda*, *V. lutea* and *V. tricolor*, are increased by seed sown in autumn. Most violas are easy from summer cuttings and several can be divided in autumn.

WAHLENBERGIA Propagation is effected by seed sown in February or by cuttings secured from June to August. The seed of *W. hederacea* should be sown on finely chopped fresh sphagnum moss. This species is also increased by division in March.

ZAUSCHNERIA (Californian fuchsia) Easily propagated by cuttings taken in summer or autumn. These should preferably be wintered indoors.

18
Increasing Bulbous Plants

Bulbous plants are taken here to comprise plants which produce bulbs, corms or tubers. Most of these plants are increased by natural methods such as seed, offsets and bulbils. Artificial methods include the taking of scales, as with lilies, and the 'scooping' of hyacinth bulbs. Fig. 66 shows the basic technique used in planting bulbs.

How the different kinds are increased
ACIDANTHERA Increase by seed in heat or by natural division.
ALLIUM Lift and divide the clumps when dormant and replant in autumn. Seed may be sown when ripe.
AMARYLLIS The bulbs of *A. belladonna* resent disturbance and should not as a rule be lifted more than once in five years. This should be done in July and offsets are then secured and replanted. A warm situation is necessary.
ANEMONE (windflower) Sow seed as soon as ripe in July or August in open beds. These will flower in spring. The tubers also increase naturally and should also be lifted annually in June or July. Dry them off, divide and replant in the autumn. Seed sown in spring often flowers in autumn.
ANTHERICUM (St. Bernard's lily) Seed will germinate in a cold frame if sown just after ripening. Division is another method.
ANTHOLYZA Seeds are sown in gentle heat in spring. Lift and divide bulbs in August, dry and replant in October. These require a sunny border.
ARISAEMA Increase by division of the tuberous roots as growth starts.

BABIANA (baboon root) Propagated by offsets which are grown on in boxes or planted out in light soil. Seed will germinate in gentle heat.
BEGONIA Tuberous-rooted begonias are easily raised from seed sown in January in heat. The seedlings are grown in a light compost and are potted on as they grow. They are gradually hardened off for planting outside in May. The tubers may be stored over winter in a slightly heated greenhouse and are sometimes cut into several pieces at planting time, each with a bud attached.
BRODIAEA (Californian hyacinth) Lift, divide and replant in autumn.
BULBOCODIUM (spring meadow saffron) Lift when dormant and plant offsets in autumn.
CALOCHORTUS (star tulip) Increased by seeds, offsets and bulbils. Sow seeds in pans in a cool house or frame when ripe or in early spring. Divide and replant the bulbs in autumn.
CAMASSIA (quamash) Readily increased from seed sown outside when ripe. Also by offsets in autumn or spring.
CARDIOCRINUM (giant lily) Offsets can be removed in autumn and grown on to flowering size. Seeds can be sown in a frame in autumn.
CHIONODOXA (glory-of-the-snow) Increase by seed is the best method, and this should be sown in cold frames. The seedlings should be left undisturbed for two or three years.
CLIVIA All species may be propagated from offsets which are potted separately in 12 cm (5 in) pots when repotting stock in February or March.
COLCHICUM (meadow saffron) Lift the bulbs in June, grade into sizes and replant. Seed may be sown in open beds or in pans or boxes stood in a

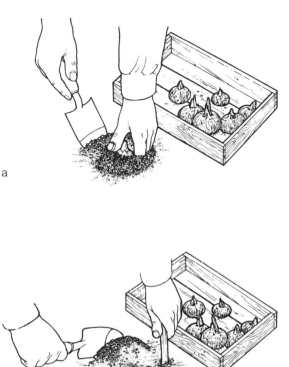

Fig. 66. Bulbs are usually planted in their dormant season to twice their own depth. Take out a hole with a trowel (*a*), insert the bulb (*b*) and replace the soil firmly. Mark the site occupied by the bulbs to prevent inadvertent disturbance (*c*).

cool greenhouse or frame in August or September. Seedlings take five years to reach flowering age.

CRINUM Seeds of the hardy species are sown in spring in a heated greenhouse. Lift and divide in March. Sometimes the bulbs are lifted in autumn, stored over winter and replanted in the spring. A warm site is necessary.

CROCOSMIA The popular montbretias of gardens have originated from *C. aurea*. The corms are usually lifted, divided and replanted in autumn. In cold districts, however, they should be stored over winter and planted in spring.

CROCUS These require a light, dry soil. They are usually lifted annually for propagation purposes when the foliage dies down, which varies with different species. The corms are dried, graded and replanted.

CYCLAMEN (hardy species) Seed collected and sown when ripe will germinate in cold frames or in the open.

ENDYMION (bluebell) Divide in autumn. Seeds, if available, can be sown in peaty soil in spring.

ERANTHIS (winter aconite) Sow seed in open beds in September and leave the seedlings undisturbed for two or three years. Divide the tubers in October or November.

ERYTHRONIUM (dog's tooth violet) Increased by offsets planted in August.

EUCOMIS Remove and replant offsets in autumn.

EUCHARIS Remove and pot up offsets in summer; seeds can be sown in sandy compost in a temperature of 27°C (80°F) in winter or spring.

FRITILLARIA (fritillary) Secure offsets in August and replant. Sow seeds when ripe in a cold frame.

GALANTHUS (snowdrop) Divide and replant immediately after flowering. Sow seed when ripe in boxes or pans stood in the open.

GALTONIA (spire lily) Secure offsets from lifted plants in autumn and replant. Sow seed in a cold frame in spring or summer.

GLADIOLUS Large quantities are raised commercially. The corms are planted in rows 30 cm (12 in) apart with 15 cm (6 in) between each corm, or they may be planted in beds 15 to 23 cm (6 to 9 in) apart each way. Depth of planting is 8 to 12 cm (3 to 5 in). The grandiflorus and primulinus types are planted between February and April. The colvillei and early flowering kinds are winter-hardy and are planted in October. The latter are lifted in August but the later flowering kinds are left until October and stored over winter.

Remove most of the foliage and dry thoroughly in a warm house after lifting; then remove the stem bases, separate and clean the corms. Store in a dry, frost-proof shed. Very tiny corms, called spawn, should be stored in sand and are sown in drills in the spring. Corms are sometimes slashed with a knife at their bases to induce the production of spawn.

The corms are graded into sizes. Those of the grandiflorus type above 9 cm ($3\frac{1}{2}$ in) in circumference are used for flower production, and slightly smaller sizes of the primulinus type for the same purpose. Smaller sizes are replanted to increase their size.

HABRANTHUS Divide the bulbs when lifting or repotting.

HAEMANTHUS Remove offsets in spring or autumn and repot individually. Seeds can be sown in a temperature of 18°C (65°F) in spring.

HYACINTHUS (hyacinth) The hyacinth rarely produces offsets naturally, so artificial measures are necessary. Mature, but not old, bulbs with thick, fleshy scales are selected when ripe, usually towards the end of August. There are two methods of treatment. The first is called cutting and consists of making cuts in the form of a star right across the base of the bulb. The second method is termed scooping, and is generally preferred to cutting. It involves the very careful cutting out of the whole basal plate of the bulb with a curved scalpel or sharpened spoon. Only the extreme end of each scale must be cut.

After treatment the 'cut' bulbs are packed in a box and are surrounded with dry sand where they are left for two to three weeks to callus. 'Scooped' bulbs are dipped in fungicide powder and are placed scooped side upwards in wire-bottomed trays for callusing. Dry conditions are essential.

When callusing is complete the treated bulbs are incubated in a temperature of 22°C (72°F) with moist conditions, maintained by sprinkling water on the floor. Small bulblets develop on the cut surfaces and, when roots are seen to be developing on these, the bulbs are planted in pots (bulblets upwards), usually in October or November, so that the bulblets are just buried. They may also be planted outdoors.

Mulching with straw protects the bulbs from frost. In the following July the bulbs are lifted, dried and replanted in September. This is repeated for four to five years when the bulbs reach flowering size. Seed is another method of propag-

ation, and should be sown in the autumn in a cold frame.

IPHEION (spring star flower) Divide overcrowded clumps in spring or autumn.

IRIS Spanish, English and Dutch irises are lifted in July or August and graded into flowering bulbs and offsets. Replanting is usually carried out in August or September. Seed may be sown in a cold frame as soon as ripe. Early flowering irises such as the reticulata type are lifted and divided when dormant.

IXIA (African corn lily) Lift bulbs when the foliage dies down, dry off and store in a warm room. Replant September to January. Ixias are hardy only in mild districts.

LEUCOJUM (snowflake) Increased by offsets lifted and replanted in September or October.

LILIUM (lily) There are several methods of increase as below:

1. *Seed*. This method is preferable because there is less risk of transmitting virus diseases. Seed may be sown in spring or autumn, whichever is most convenient. Sow in cold frames or in boxes placed in slightly heated greenhouses. Light or peaty composts are quite suitable but should not be over rich in fertilizers. Some species such as *L. regale*, *L. formosanum* and many hybrids germinate quickly and easily, other kinds such as *L. giganteum* may take up to two years. With other species like *L. monadelphum*, *L. canadense* and *L. superbum* the seed germinates and a small bulb is formed but top growth does not appear until the second year after sowing.

For large-scale production, seed of many species, such as *L. regale*, may be sown in spring in light, well-drained soil in the open. Sow in shallow drills 30 cm (12 in) apart. If sown thinly the seedlings may be left to develop for two or three years.

Whatever the method the seedlings should be grown cool during the summer, trays or pans being stood outside in a shady position. The little bulbs may be left in the containers over winter with the compost kept only slightly moist. In spring they may be planted out for growing at about 8 or 10 cm (3 or 4 in) apart. Some lilies will flower in their second season when raised from seed. Examples are *L. tenuifolium*, *L. formosanum* and *L. regale*.

2. *Scales*. All true bulbs consist mainly of fleshy leaves or scales. With lilies these can be used for propagation (Plate 7a). To do this the scales are

carefully detached from the bulbs and inserted base downwards in a propagating medium or in light sandy soil. The scales are best secured when the plants are just past their flowering peak. They should be planted in beds or in containers 6 cm (2½ in) deep in rows 15 cm (6 in) apart with 2.5 cm (1 in) between the scales. Afterwards a further covering of sphagnum moss may be given. Moderate bottom heat will promote the formation of bulbils, but the compost must be kept fairly dry for the first four to six weeks otherwise the scales may rot. Bulbils are formed on the scales within a period of four to eight weeks. These are removed and treated similarly to bulbs raised from seed.

Another method of scale propagation is to place the scales in polythene bags with a mixture of moist peat and sand. The bags are then tied tightly and kept in a warm place until bulbils are formed on the scales.

3. *Bulbils.* Certain species of Lilium produce underground bulbils naturally. Examples are *L. longiflorum*, *L. auratum* and *L. speciosum*. These are usually removed in late summer and autumn and may be planted when convenient. Other species such as *L. tigrinum* (tiger lily) produce bulbils on their stems above ground which are also suitable for propagation (Plate 7b). Lastly, there are lily species like *L. dauricum* and *L. candidum* (madonna lily) which can also be induced to form bulbils by pulling up the flowering stems and replanting them in a trench about 10 cm (4 in) deep. Bulbils form on the stem bases and can then be removed for planting.

4. *Division.* This should be carried out when bulb groups become overcrowded. It consists of lifting and carefully breaking up the clumps into individual bulbs for replanting.

MUSCARI (grape hyacinth) Lift and replant the bulbs in autumn. Seed may be sown in a cold frame in spring or in a heated greenhouse in autumn.

NARCISSUS (daffodil) These have a fairly rapid natural rate of increase, and for bulb production are lifted annually, usually in July and August when the foliage has died down. They are then thoroughly dried, preferably in a well-ventilated shed, and are afterwards cleaned and graded. In grading four main types are selected, namely mother bulbs, double-nosed bulbs, round bulbs and offsets. When planted, mother bulbs usually produce a number of offsets, double-nosed develop into mother bulbs, round bulbs become double-nosed and offsets form round bulbs, and so the propagation sequence continues. The bulbs are stored until planting time in a cool, well-ventilated store.

Planting is usually carried out from August until October, the earlier the better. On a field scale the bulbs are planted in furrows made by a plough. The rows are usually 23 cm (9 in) apart and the bulbs spaced 2.5 cm (1 in) or so apart, the planting depth being 8 to 12 cm (3 to 5 in). Planting in beds is more suitable for gardens, and in this case the rows are also spaced about 23 cm (9 in) apart across the bed.

During the growing season weeds must be kept down, diseased plants should be removed and at flowering time any rogues should be pulled out. Sometimes the crop is deblossomed to encourage the production of better bulbs.

NERINE (Guernsey lily) These increase naturally but rather slowly and should be left undisturbed for as long as possible. Once in three years is often enough to lift groups for dividing and replanting or repotting. This is best done in August. Seeds provide another method of increase and should be sown when ripe in pans or trays kept in moderate heat.

NOMOCHARIS Usually raised from seed which is sown thinly in pans or boxes in February under cool conditions. When the seedlings are about 8 cm (3 in) high they are transplanted into a cold frame and grown under cover until October. In January the bulbs may again be transplanted into deep boxes and grown in moderate heat until April when they are planted in the open in light soil. Flowering occurs in from three to four years after sowing.

ORNITHOGALUM (star of Bethlehem) Divide the bulbs in autumn.

OXALIS (wood sorrel) Easily increased from seed germinated in a temperature of from 16 to 18°C (60 to 65°F) in spring. Division when repotting is another method.

PARADISEA (St. Bruno lily) Divide in spring or autumn; sow seeds in a frame in spring.

PUSCHKINIA (striped squill) Lift and secure offsets and replant in October. Seed is sown in pans, placed in a frame in August or September.

RANUNCULUS (buttercup) The tuberous-rooted species are increased by division and some by autumn-sown seed.

RHODOHYPOXIS Divide established plants in spring; sow seeds in a frame in spring.

ROMULEA Increased by seed and offsets.

SCHIZOSTYLIS (Kaffir lily) Readily increased by division in spring. Seed is also used when available.

SCILLA (squill, bluebell) Lift and divide the bulbs in autumn when they are becoming overcrowded.

SPARAXIS Lift and replant in the autumn. Grow in a sunny border.

STERNBERGIA (lily of the field) Lift and divide in autumn. Requires a warm position.

TECOPHILAEA (Chilean crocus) Remove and plant up offsets individually in summer.

TIGRIDIA (tiger flower) Lift the bulbs in October and store in a dry, warm house over winter. Replant in March. Sow seed in spring. Grow in a warm border.

TRILLIUM (American wood lily) Usually propagated by division but seed sown under cool conditions is another method.

TRITONIA Propagate by seed or division.

TULIPA Tulips are lifted annually, usually about the end of June. The bulbs should preferably be dried in a heated store and are afterwards cleaned and graded. They should be handled carefully. Bulbs over 9 cm ($3\frac{1}{2}$ in) in circumference are usually planted for flowering, but up to this size they are used as planting stock. Mother bulbs which are usually over 14 cm ($5\frac{1}{2}$ in) may also be planted for stock and give rise to a number of offsets.

Planting is usually carried out in October. The bulbs should be set 8 to 10 cm (3 to 4 in) deep to the bulb base. The rows are spaced 23 cm (9 in) apart and about a bulb width is left between the bulbs. During the growing season plants showing signs of disease are removed at once and the beds are kept weeded. Plants grown for bulb production should have their flowers cut off immediately they open, but the stalks are left.

VALLOTA (Scarborough lily) Sow seeds in a temperature of 16°C (60°F) in spring; remove offsets in autumn and pot up individually.

WATSONIA (bugle lily) Offsets and seed are the methods of propagation.

ZEPHYRANTHES (zephyr flower) Lift and divide the bulbs in autumn. Grow in warm borders.

19
Raising Annuals and Biennials

Annual flowering plants grown in the garden are divided into two distinct groups; namely hardy annuals and half-hardy annuals. The former are usually sown in the open where they are to flower, while the latter are normally started in a heated greenhouse for planting outside when the weather is congenial.

Hardy annuals

In mild districts hardy annuals may be sown in early autumn (early September). This may result in early and better flowering than with spring sowing.

In this respect some kinds are hardier than others. Thus the following annuals have a relatively good chance of winter survival when sown in autumn: *Centaurea cyanus* (cornflower) *Iberis* sp. (candytuft), *Delphinium ajacis* (larkspur), *Nigella damascena* (love-in-a-mist) and *Calendula officinalis* (pot marigold). Examples of less hardy annuals are: chrysanthemums, *Clarkia elegans, Godetia grandiflora, Gypsophila elegans, Centaurea moschata* (sweet sultan). In cold districts, however, autumn sowing is not likely to be successful, as the young seedlings are usually killed during the winter. Spring sowing is, therefore, preferable in most parts of the country, and early April is the best time.

The soil for annual seed should be in extra good condition. For spring sowing it should be turned over in autumn and left exposed to the weather. Fertility should be ensured by the application of lime (if necessary) and fertilizers. Phosphates are particularly important for seedlings, and a dusting of superphosphate raked in before sowing is usually beneficial. Heavy soils should be lightened by the addition of such materials as leafmould, peat and sand. Prior to sowing, the surface should be broken down fine and made level.

Annual flowers are usually grown for two purposes: (*a*) for garden decoration, (*b*) for cut flowers. In the former case a border may be devoted entirely to annuals.

Normally the seed is sown evenly and thinly broadcast. To ensure even distribution of seed, the technique of marking out relatively small areas with sand is recommended (Fig. 67). Thick sowing may induce damping-off disease. The seed is covered by carefully raking, but it is usually beneficial in addition to scatter some dry soil over the sown area. This is of particular importance on land inclined to be wet and sticky. Large seed should be sown in drills drawn at regular intervals across the patches.

Annuals to be grown for cut flowers can be sown in beds 1.25 to 1.5 m (4 to 5 ft) wide with alleyways 60 cm (2 ft) wide between the beds. Drills are made across the beds 30 cm (1 ft) apart. This treatment is suitable for such annuals as marigolds, nigella and clarkia. The taller-growing species, however, like larkspur and cornflowers, are usually grown in single rows 45 cm (18 in) apart.

When the plants are well through the soil they must be thinned out to 15 to 23 cm (6 to 9 in) apart. September-sown crops are thinned out either in late autumn or left until early spring. Thinnings (if lifted carefully with plenty of soil around their roots) may be used to fill in gaps. Weeds must be kept down in the early stages.

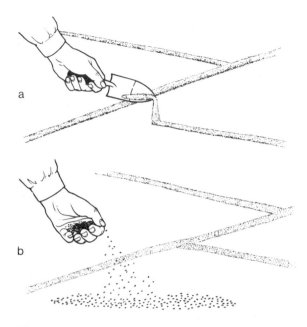

Fig. 67. Broadcast sowing. (*a*) Use sand to mark out the 'drifts' to be sown. (*b*) Sow seed thinly and evenly within its own defined area.

Hardy annuals include a wide range of plants, and a selection of these is given below:

Adonis aestivalis (pheasant's eye)
Agrostemma 'Milas' (corncockle)
Alyssum maritimum (madwort)
Anchusa capensis
Asperula orientalis
Atriplex hortensis
Calendula officinalis (pot marigold)
Centaurea cyanus (cornflower)
Centaurea moschata (sweet sultan)
Centaurium venustum
Chrysanthemum coronarium
Chrysanthemum segetum (corn marigold)
Cladanthus arabicus
Clarkia elegans
Collinsia bicolor
Convolvulus tricolor
Coreopsis drummondii (tickseed)
Coreopsis tinctoria (tickseed)
Cosmos bipinnatus (Mexican aster)
Delphinium ajacis (larkspur)
Dianthus chinensis (Chinese pink)
Dimorphotheca aurantiaca
Dimorphotheca pluvialis

Eschscholzia californica (Californian poppy and its varieties)
Felicia bergeriana (kingfisher daisy)
Gaillardia pulchella (blanket flower) and its varieties
Gilia tricolor
Godetia grandiflora and its varieties
Godetia amoena and its varieties
Gypsophila elegans (chalk plant)
Helianthus annuus (sunflower) and its varieties
Helianthus debilis
Helichrysum sp. (everlasting flower)
Helipterum (immortelle flower)
Humulus japonicus (hop)
Iberis amara (candytuft) and its varieties
Ipomoea tricolor (morning glory)
Lathyrus odoratus (sweet pea)
Lavatera trimestris (mallow)
Layia elegans (tidy tips)
Limnanthes douglasii (poached egg flower)
Linaria maroccana (toadflax)
Linum grandiflorum (flax)
Lupinus hartwegii (lupin)
Lychnis coronaria (campion)
Malcomia maritima (Virginian stock)
Malope trifida
Matricaria maritima (double mayweed)
Medicago echinus (calvary clover)
Mentzelia lindleyi (bartonia)
Mimulus cupreus and its varieties
Moluccella (bells of Ireland)
Nemophila menziesii (baby blue-eyes)
Nicandra physaloides (shoo-fly plant)
Nigella damascena (love-in-a-mist)
Papaver (poppy—annual species)
Phacelia campanularia
Reseda odorata (mignonette)
Salvia horminum
Saponaria calabrica (soapwort)
Scabiosa atropurpurea (sweet scabious)
Senecio elegans
Silene armeria and *S. pendula*
Tagetes erecta (African marigold)
Tagetes patula (French marigold)
Tagetes tenuifolia var. *pumila*
Tropaeolum (nasturtium)
Ursinia versicolor
Verbena × *hybrida* (vervain)

Half-hardy annuals
Half-hardy annuals are widely used for giving a summer display in gardens. Huge quantities of

the most popular sorts are raised every year in nurseries. Seed of the various kinds is sown from January to March in heated greenhouses. Early sowing is necessary with such species as antirrhinums, but many other kinds develop more quickly and may be sown later. Sometimes antirrhinums are sown in autumn.

Seed trays or boxes are generally used for raising annuals (Fig. 68). They should be filled with John Innes or a peat-based seed compost and made firm and level on the surface. The boxes should then be well watered and left to drain before sowing. Scatter the seed thinly with the fingers and cover lightly with the same compost sieved over the surface. Dust-fine seeds should be left uncovered. Afterwards the trays are placed on the staging and covered with glass and paper. A temperature of 13 to 16°C (55 to 60°F) is suitable.

When the seedlings appear the glass and paper coverings are removed at once and the seedlings watered as required, using a fine rose. As soon as possible the seedlings should be pricked out into other boxes, this time using a potting compost. The spacing is such that 35 seedlings (7×5) are planted in a standard plastic seed tray.

After pricking out, the seedlings are kept relatively shaded and warm for a few days to help them to recover from the move. Some shading may also be necessary. They should be watered sufficiently just to keep the soil nicely moist, and the same temperature maintained. Gradually, as the plants develop, more ventilation is given and in March or early April the boxes are transferred to cold frames. Here the hardening-off process continues until finally the lights are removed a week or two in advance of planting out.

A list of popular half-hardy annuals is given below:
Ageratum houstonianum and its varieties
Alonsoa warscewiczii
Amaranthus caudatus (love lies bleeding)
Antirrhinum majus (snapdragon) and its varieties
Arctotis grandis
Brachycome iberidifolia (swan river daisy)
Browallia demissa and *B. viscosa*
Callistephus chinensis (China aster)
Celosia cristata (cockscomb) and its varieties
Cleome spinosa (spider flower) and its varieties
Gazania splendens (treasure flower)

Impatiens (balsam) and its varieties
Ipomoea (morning glory)
Kochia scoparia (summer cypress)
Lobelia erinus and its varieties
Matthiola incana (ten-week stock)
Mesembryanthemum (Livingstone daisy)
Nemesia strumosa and its varieties
Nicotiana (tobacco plant)
Petunia varieties
Phlox drummondii
Salvia splendens and its varieties
Salpiglossis sinuata
Schizanthus × *wisetonensis* and its varieties
Ursinia anethoides
Venidium fastuosum
Verbena—annual species
Zinnia elegans

Biennials

Biennials or species treated as such comprise a very useful group of plants for spring and early summer flowering. Seed is usually sown in the open from May until August. Early sowing, however, gives the plants a chance of becoming well established before the winter. Sow in drills drawn 30 cm (12 in) apart. If the soil is dry the open drills should be well watered before sowing.

When the seedlings are large enough to handle they are pricked out in rows or in beds 15 to 23 cm (6 to 9 in) apart. In October the plants are usually set in their flowering quarters.

Below is given a list of popular plants treated in this way, with additional notes on their propagation.

ALTHAEA (hollyhock) Sow seeds of *A. rosea* varieties in drills 30 cm (12 in) apart in June, choosing a warm, sunny border. Thin the seedlings out to 15 cm (6 in) apart and transplant to flowering quarters in September.

ANCHUSA Sow *A. capensis* outdoors in April or May for flowering the following year. May require some protection in winter.

CAMPANULA Sow Canterbury bells (*C. medium*) and *C. pyramidalis* outdoors April to June. Transplant seedlings 15 cm (6 in) apart in beds and plant in permanent positions in October.

CELSIA Sow *C. cretica* in the open in April or May in a sunny position. Thin out or transplant the seedlings.

CHEIRANTHUS (wallflower) This excellent group of spring-flowering plants prefers a light, well-

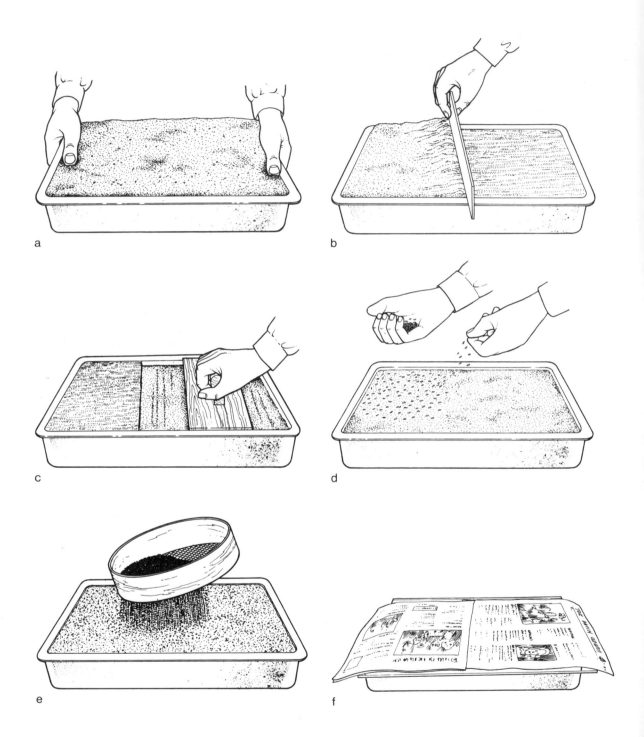

Fig. 68. Sowing seed in a seed tray. (*a*) Slightly overfill tray with seed compost and tap tray to settle the compost. (*b*) Remove excess mix with a straight edge. (*c*) Firm gently with a rectangular presser-board. (*d*) Sow seeds thinly and as evenly as possible. (*e*) Cover the seeds with sieved seed compost. (*f*) Place a sheet of glass covered with newspaper over tray: remove both on germination of seeds.

drained soil. Sow in shallow drills in May and transplant the seedlings 15 cm (6 in) apart in beds when they are 5 to 8 cm (2 to 3 in) high. Set the seedlings in their flowering quarters in October. The popular Siberian wallflower (*C. allionii*) is treated in the same manner. Wallflower seed is subject to a condition known as thermo-dormancy. This is due to exceptionally hot sunny weather in, say, late May or early June. Under such circumstances, if the site is fully exposed to the sun, germination may be poor or even fail completely. Thermo-dormancy is likely to occur only in the sunnier, warmer parts of the country and in such districts it may be wise to sow wall-flower seed in a semi-shaded situation.

CYNOGLOSSUM Sow the seed of *C. amabile* in June and thin out to 23 cm (9 in) apart or transplant in September or April.

DIANTHUS Sow *D. barbatus* (sweet William) out-doors in April or May, prick out in rows and plant in their flowering quarters in October.

DIGITALIS (foxglove) Sow in late May or early June and transplant in September about 23 cm (9 in) apart.

ECHIUM (biennial species) Sow in May for flower-ing the following season in a sheltered position.

EUPHORBIA (spurge) Sow *E. lathyrus* in late May and transplant 20 cm (8 in) apart in September.

HESPERIS (sweet rocket) *H. matronalis* is usually sown at the end of May and transplanted 30 cm (12 in) apart in late September.

LUNARIA (honesty) Sow *L. annua* (syn. *L. biennis*) in April in a sunny position, thin out or transplant to permanent position in August or September.

MATTHIOLA (Brompton stock) *M. incana* is usu-ally treated as a biennial. Sow in a cold frame in June or July. Transplant the seedlings, when they are 2.5 cm (1 in) or so high, to their flowering quarters in a sheltered position. In cold districts overwinter in a cold frame.

MYOSOTIS (forget-me-not) Sow outdoors in April, May or June, prick out and transplant to flower-ing position in October.

OENOTHERA (evening primrose) Sow *O. biennis* in May and transplant 23 cm (9 in) apart in Sep-tember.

PAPAVER (Iceland poppy) Sow the seed of *P. nudicaule* in May where they are to remain and thin out to 23 cm (9 in) apart.

PRIMULA hybrids (polyanthus) These are true perennials and may be increased by division in autumn. They give better results, however, when treated as biennials. In order to have large flower-ing plants for the following year the seed should be sown in February or March in gentle heat; prick out and plant outside later. They can also be sown outside in April or May but the plants will be smaller than those sown earlier.

VERBASCUM Sow *V. bombyciferum* and *V. thapsus* in June and transplant 15 cm (6 in) apart in September. Transplant to flowering site in spring.

VIOLA (violas and pansies) Viola and pansy var-ieties are propagated from summer cuttings which are usually inserted in a cold frame. Sev-eral good strains of violas in distinct colours come true from seed and excellent strains of pansies in mixed colours are also available. Normally the seed is sown in boxes in March or April and germinated in a slightly heated greenhouse or cold frame. Prick out into other boxes and grow in a frame until planting out.

145

20 The Propagation of Hardy Fruit

Berried fruits

All the berried fruits are grown on their own roots and as a general rule are easy to propagate. With blackcurrants, raspberries and strawberries, however, the main difficulty is to keep the plants healthy. These fruits are subject to certain virus diseases which cause a gradual deterioration of the plants and a reduction in crop yield. It is, therefore, most important to propagate only from healthy stocks. Methods of combating virus diseases are discussed in Chapter 23, but it is usually wise for the amateur to get expert advice before attempting to increase his own plants.

BLACKCURRANTS (*Ribes nigrum*) The usual method of increase is by hardwood cuttings, which should preferably be secured immediately after leaf-fall. Select firm, well-ripened shoots of the current season's growth, but if these are hard to find, two- or three-year-old wood may be used.

It is best for the amateur to plant the cuttings the same day as they are secured. In nurseries, however, this may not be practicable and very often, after preparation, the cuttings are tied in bundles and laid with their stem bases in moist soil or sand in a shady position. Here they will form a callus and may be left until spring.

Land for cuttings should be free from weeds, and the soil well manured and cultivated. There are two methods of planting. One is to take out a trench and, after inserting the cuttings, replace the soil and tread it in firmly. The other method simply consists of pushing the stems into loose, freshly turned-over soil, which is then trodden in firmly around the cuttings.

Commercially, blackcurrant cuttings are planted in rows 68 to 75 cm (27 to 30 in) apart with 15 cm (6 in) allowed between the cuttings. In private gardens the rows may be 45 cm (18 in) apart. The depth of planting is normally 18 to 20 cm (7 to 8 in) leaving about 8 cm (3 in) of the cutting above soil level. When growth begins the topmost buds produce shoots which arise at or just below the soil surface, forming a stool type of bush without a main stem or 'leg'.

Under good conditions growth is fairly rapid and at the end of the first season shoots 60 to 100 cm (2 to 3 ft) long may be produced. Shortage of moisture is a likely cause of poor results and it is often recommended to mulch the cuttings in the early spring to help in retaining moisture. Hop manure, pulverized bark, leafmould, and coarse peat are ideal for this purpose.

In autumn the plants may be transplanted either to their permanent position, or in nursery rows 75 cm (2½ ft) apart with 30 cm (12 in) left between the plants. After transplanting, the young shoots should be cut back to within a bud or two of their point of origin. These young shoots may be used for propagation.

Another method of increasing blackcurrants is from softwood cuttings. These are usually secured in May from the immature leafy shoot tips, and are cut off below a leaf so that they are about 8 cm (3 in) long. The cuttings are inserted firmly with a dibber in sandy compost in a cold frame about 15 cm (6 in) apart.

Shade and warm conditions must be provided and the compost watered until the cuttings have rooted, which is usually in quite a short time. As the cuttings become established the shade is gradually removed and more ventilation is given until finally the lights are removed. In the autumn the rooted cuttings are planted outside in nursery rows and have their young growths cut hard back. Subsequent treatment is the same as for hardwood cuttings.

Blackcurrants can also be easily propagated by mound layering. The method is to cut back a plant intended for propagation to near ground level. The young shoots that arise from the base are kept earthed up as they grow and will produce roots below the soil. These rooted shoots may be pulled off and transplanted in autumn.

RED AND WHITE CURRANTS (*Ribes sativum*) Normally these are also increased from hardwood cuttings, firm one-year-old stems being preferable. The cuttings should be taken in autumn and are made 30 to 35 cm (12 to 14 in) long. All the buds with the exception of three or four at the tip are neatly trimmed off with a sharp knife. This means that shoots will be produced only from near the tip of the cutting, to eventually form the head of the bush. Unlike blackcurrants, the red and white types are best grown with a clean stem or 'leg'.

The cuttings may be planted the same distance apart as advised for blackcurrants, but the depth of planting should not be more than 15 cm (6 in). Best results are secured on good, rich soil and mulching in the spring between the rows to conserve moisture is beneficial.

The young plants are usually transplanted in autumn and each young shoot is cut back to within two or three buds of its origin. In subsequent years pruning is adjusted to build up a permanent framework of branches. Redcurrants may also be propagated from softwood cuttings in exactly the same manner as blackcurrants. As the young plants grow, however, they will have to be trained into the normal shape recommended for this fruit. Layering is another possible method, but plants for this purpose should be grown on the stool principle.

GOOSEBERRIES (*Ribes grossularia*) The method of increasing gooseberries from hardwood cuttings is almost exactly the same as for red and white currants. Gooseberries should also have a clean stem or 'leg'.

Firm, well-ripened stems of the current season's growth are essential, and coarse, soft, young shoots should be avoided. It is often difficult to secure suitable cuttings from fruit bushes. The best type is produced from stool beds specially grown for this purpose. Here the bushes are cut down annually and if given good treatment produce plenty of young shoots each year. The cuttings should preferably be secured in autumn and are best cut below a node or taken with a heel. Planting should take place at once, but if this is not possible the cuttings may be laid in the soil for planting later.

Gooseberries may also be increased by layering, but with this method it is difficult to secure a shapely bush. Sometimes suckers arise naturally and may be transpanted and encouraged to develop into strong plants.

RASPBERRIES (*Rubus idaeus*) These may be said to propagate themselves with very little assistance from the gardener. Very often indeed the numbers of new canes that arise naturally alongside the permanent row are an embarrassment to those who require fruit only. These young canes may be transplanted in autumn or winter and afterwards are usually cut down to within about 23 cm (9 in) of ground level. Plants secured in this manner are quite suitable provided they are not affected with the troublesome virus disease called mosaic.

The method of raising raspberries commercially is to establish what is called a cane nursery. This is done by planting healthy virus-tested canes in rows 2 m (6 ft) apart with 60 cm (2 ft) between the plants. Light soil is most suitable as it allows the roots to spread freely and throw up an abundance of suckers. Reasonable freedom from weeds is essential but weeds may be prevented from growing if the soil is treated with the herbicide simazine when it is in a weed-free condition.

The canes are not allowed to flower or produce fruit, but are cut down periodically. Should any flowers appear these are removed to prevent seed production and, consequently, unwanted seedlings appearing in the rows.

Under these conditions masses of young canes very often arise, and are dug up for sale as required. The bed usually remains in production for three or four years when it is cleared and a fresh start made on a new site.

A cane nursery should be well isolated from fruiting plantations to reduce the risk of mosaic infection. The cane beds are inspected frequently and any plants showing signs of disease are

147

removed, together with their neighbours, immediately. In this way a cane nursery can be kept in a much healthier condition than is possible with fruiting plantations. Raspberries may also be increased from root cuttings.

BLACKBERRIES AND RELATED HYBRIDS (*Rubus ursinus* and varieties) Blackberries, loganberries and similar cane fruits are normally propagated from tip layers. With ordinary plants grown for fruit this may be done in August. Light soil well supplied with organic matter is most suitable. The usual method is to make a slit in the soil with the spade or trowel and insert the tips of young canes in this to a depth of about 15 cm (6 in). The soil is then trodden firmly around the shoot.

The layered tips quickly take root and a new shoot is produced from below the soil. The new plants are usually severed from their 'parents' and transplanted the following spring.

The commercial method of tip layering blackberries is to establish a stool bed on clean, light soil, planting the canes 2 m (6 ft) square in November or February. In spring the plants are cut back to ground level and the soil is kept cultivated and free from weeds.

Young shoots arise from the base of the plants, and when they are about 75 cm (2½ ft) long their tips are pinched out, with the object of encouraging the production of laterals. Layering is usually done from mid-June until the end of July, the tips of the sideshoots or laterals being inserted with a dibber, after removing the topmost leaves. It is essential to tread in firmly.

By the autumn the layers are usually well rooted and may then be severed and transplanted in nursery beds, and should be staked. A stool bed lasts for about four years, but after that it is advisable to secure a fresh site. Cane tips may, of course, be layered into pots or other receptacles, and this is a convenient method for the amateur.

Other methods of propagating these fruits are by ordinary layering, by root cuttings and by leaf-bud cuttings (Chapter 11).

STRAWBERRIES (*Fragaria vesca* varieties) Runners are the natural, efficient and speedy method of increase. In fruiting plantations after the crop has been cleared, and the bed freed from weeds, the runners may be pegged down to enable them to root more quickly. They are usually ready for transplanting in September and October but it is better if they can be had in August. Runners planted in August often produce a fair crop the

following season, whereas this is unlikely with later planting.

As with raspberries, commercial strawberry propagation involves measures to check the spread of virus diseases, which can cause rapid deterioration of the plants. This is usually achieved by what is called the isolated block method which facilitates inspection and the removal of diseased plants. The plants are set 1 m (3 ft) apart and each group of four plants have all their runners trained towards one another so that the groups form separate blocks. Should a diseased plant be observed in a block the whole square of plants and their runners is destroyed.

Beds of this type are not allowed to fruit, and as all the energies of the plants are concentrated on runner production large numbers of these usually arise.

Sometimes a limited number of runners is pegged into small pots filled with rich compost which are sunk in the ground. This method is occasionally used for the production of early plants which are planted not later than August for forcing in frames or cloches the following spring. Pot-layered plants may also be used for forcing in greenhouses.

Runners should always be lifted carefully to avoid root damage and care should be taken that the roots are not exposed to sun or drying winds before being planted. Division is another method of propagation which may be used if runners are scarce. It consists of splitting up large plants into several crowns.

Alpine and perpetual strawberries are raised from seed which is usually sown in pans in the autumn, and left outside over the winter. In early spring the pans are placed in an unheated greenhouse or frame where the seed usually germinates freely. The seedlings are pricked out when large enough and later are planted in the open.

Tree fruits

Tree fruits are usually increased by budding or grafting a scion (which eventually becomes the tree head) on to a rootstock which supplies the root system. The stock affects rate of growth, height and spread of the branches, the age when the tree begins to fruit, time of blossoming and the colour and size of the fruit.

Apple rootstocks

These were first classified at the East Malling

Research Station and the earliest selections were given the prefix 'Malling', each one being distinguished by a number always written in the form of a Roman numeral, e.g. Malling (or M) IX. Later a further range of rootstocks was raised by East Malling working in conjunction with the John Innes Horticultural Institute then at Merton. These stocks were given the prefix Malling Merton (or MM) followed by a number for each stock, e.g. MM 106. A third group was raised at East Malling and is again distinguished by the prefix Malling (or M) followed by a number in ordinary figures, e.g. M 26. It has now been decided that in future all stock numbers will be in ordinary (Arabic) numerals.

Few amateurs will attempt to raise their own fruit trees but those who wish to have a go should have information on the kind of stocks available and the effect these will have on the trees. Similarly those who purchase fruit trees from nurserymen should always insist on securing the name and number of the stock used and also ensure that it is suitable for the purpose. The rootstocks are:
GROUP A is described as dwarfing because such stocks produce a small compact tree which crops relatively early in its life. Hence, these stocks are ideal for small gardens. Two representatives are Malling 9 and Malling 26. The former is preferable for rich soils while M 26 is the best on poor land. Both stocks are suitable for small bush and trained stocks.

A new rootstock, M 27, is exceptionally dwarfing and encourages the fruiting varieties to crop within 3 years of grafting. It is especially useful as a stock where vigorous varieties such as Bramley's Seedling and Holstein are required in a small garden. It is also suitable as a stock for ornamental crab apples.
GROUP B (semi-dwarfing) produces trees of intermediate size suitable for the larger orchard or medium-sized garden. The two best stocks of this group are Malling 7 and Malling Merton 106. In general MM 106 is superior as it comes into cropping earlier in its life than M 7. In common with all MM stocks, too, MM 106 is resistant to the serious pest woolly aphid.
GROUP C. Stocks in this group produce large, vigorous trees rather slow to come into cropping. For the amateur, therefore, their principal merit may be to provide shade or screening in some situations apart from their advantage of bearing fruit. There are three main stocks in this group,

namely Malling 2, Malling Merton 111 and Malling Merton 104. Perhaps the best of these is MM 111 which comes into cropping relatively soon for this group and is recommended for light sandy soils. Again, it is resistant to woolly aphid.
GROUP D comprises very vigorous stocks which in the amateur's garden may be suitable only for special situations. Two typical stocks of this group are Malling 16 and Malling 25.

Pear rootstocks

Pears are usually grafted on quince stocks of which there are three types—Malling A, B and C. The A and B types have similar effects and produce medium-sized trees. Type C produces a tree which at first is as vigorous as those on A and B but after it comes into cropping vigour declines. It is therefore a good stock for small bush trees, espaliers and cordons and encourages early and heavy cropping. It is the best stock for pears in small gardens. Many varieties of pears such as William's Bon Chrétien, Marie Louise, Jargonelle, Souvenir de Congrès and Dr. Jules Guyot, when grafted on quince fails to form a successful union. This difficulty may be overcome by what is called double working which means first of all grafting a variety on the quince that unites readily with it and then grafting the desired variety on to this intermediate stem. Suitable varieties to use as intermediates are Fertility, Pitmaston Duchess and Beurre Hardy.

A three-bud scion of the required fruiting variety is usually whip and tongue grafted on to a 12 or 15 cm (5 or 6 in) scion of the intermediate variety, which is in turn whip and tongue grafted on to the stock in the usual manner.

Plum rootstocks

The principal plum rootstocks are as follows:
MYROBALAN B. Probably the best stock for producing large trees and heavy crops. It is suitable for most varieties, exceptions being Oullins' Golden Gage and Comte d'Althann's Gage. This stock tends to delay bearing of certain shy cropping varieties such as Coe's Golden Drop and Early Transparent Gage, and cannot be recommended for these. It is easily propagated by hardwood cuttings.
COMMON MUSSEL. Valued for its compatibility with all varieties. Unfortunately it produces many suckers, and some trees show a reduction in

vigour after only a few years. It is easily propagated from hardwood cuttings.

BROMPTON. Similar in vigour to Myrobalan, Brompton is compatible with all varieties but is difficult to propagate. It is a good stock for peaches, nectarines and apricots.

MARIANNA. Produces medium-sized trees which are well anchored and bear heavy crops of good-sized fruit. It is a good stock for light soil, but fails to unite with Czar, President, Oullins' Golden Gage and the damsons. Varieties that succeed on it are Victoria, Rivers Early Prolific, Pershore, Purple Pershore, Laxton's Utility, Belle de Louvain and Giant Prune. It is propagated by hardwood cuttings.

ST. JULIEN A. Perhaps the best all-round rootstock, this is a semi-dwarfing plum stock good for cropping and compatible with all varieties. It is easy to propagate from hardwood cuttings.

PERSHORE is a good plum stock for the production of small trees which will crop heavily and quickly. However, it is difficult to propagate and is not generally in use.

Some varieties of plums can be grown on their own roots; examples being Blaisden Red, Warwickshire Drooper, Pershore and Cambridge Gage, all of which may be propagated by rooted suckers.

Cherry rootstocks.

Stocks for both sweet and sour cherries, to produce vigorous trees, are secured mainly from three sources:

1. Seedling stocks imported from the Continent.

2. So-called wildings dug up in woods in this country.

3. Selected vegetatively propagated types, which are the most reliable. One of these which is being produced and distributed by the East Malling Research Station is known as Malling 12/1 (or F12/1). This stock produces well-anchored, uniform, vigorous trees and is resistant to the serious disease bacterial canker. These stocks are propagated by layering and by root cuttings. Malling 12/1 is also used for ornamental flowering cherries.

A new cherry rootstock known as Colt is also available. It is rather more dwarfing than F12/1, producing trees up to two-thirds the size of those grafted on to this rootstock. It is a hybrid between *Prunus avium* and *Prunus pseudocerasus* and should be used as a rootstock for precocious varieties such as Merton Glory if its benefit is to be fully realized. It will bring trees into bearing earlier in their life than normal, and can be easily propagated by softwood or hardwood cuttings. (Pre-formed root initials are present and make for easy rooting of hardwood cuttings without the use of hormone compounds.) Colt shows resistance to bacterial canker and cherry replant disease, produces few suckers and can be used as a stock for ornamental as well as fruiting cherries.

It is interesting to note that a wide range of fruit tree root stocks can now be propagated from hardwood cuttings as described in Chapter 11.

Budding and grafting rootstocks

Stocks of uniform size are planted in spring in rows 1 to 1.25 m (3 to 4 ft) apart, 38 cm (15 in) being allowed between the stocks. Before planting, sideshoots are removed entirely from the lower part of the stock but higher up are shortened to within 1 cm ($\frac{1}{2}$ in) of the main stem.

BUDDING. These stocks may be budded from mid-June to mid-August. In the succeeding February after budding the stocks are cut back to within 10 cm (4 in) of the buds. The piece of stem above the bud is used later as a stake for the young shoot arising from the bud. All other shoots and suckers are removed as soon as possible.

GRAFTING. Where a bud fails to grow, the stock is usually grafted in March and April, apples being left until last. For this purpose dormant scions are used, these being collected in winter and heeled-in under shaded conditions in moist soil. Each scion should be cut to three to four buds long, the unripened ends of the shoots being discarded. The stocks are cut down to within 8 to 10 cm (3 to 4 in) above the ground and the scions inserted at their apex by the whip-and-tongue method. After tying the scion is sealed with hot wax or petroleum jelly.

During the growing season the stock is kept clear of suckers. All shoots that grow on the scion should be left until they are 15 to 20 cm (6 to 8 in) long and then the strongest and best placed (usually the apical shoot) is selected to form the main stem of the new tree; the other shoots are pinched back to five or six leaves, which is better than cutting them off altogether. The graft tie is usually cut at the same time.

The young maiden shoots are usually supported with a cane or light stake. Certain trees

a

b

Plate 7. Increasing lilies by (*a*) scales and (*b*) axillary bulbils. The scales are removed from the bulbs immediately after flowering, and the bulbils can be taken from the stems of those species that form them in autumn. Both form roots in a light, sandy medium.

a

b

c

Plate 8. Increasing *Sanseveria* by division. When a plant has formed a good number of leaves (*a*) it may be entirely broken up for division, or just one or two of the shoots separated and potted up individually. Cut out one shoot with a healthy piece of fattened root (*b*), remove it from the parent (*c*) and pot it up in a 10 cm (4 in) pot of compost (*d*). Keep the new plant warm and reasonably well watered until it becomes established.

d

such as standard cherries are sometimes high worked, that is, the scion buds or grafts are inserted on stocks with tall stems—thus the scion forms the actual head of the tree only. As mentioned already certain varieties of pear fail to unite with the quince stock and must be double-worked to secure a satisfactory union.

Shaping the young trees

In the nursery the 'maiden' or one-year-old trees are trained into shape by pruning, and the method of pruning adopted depends on the type of tree required. Thus if a bush tree is wanted the maiden is cut back to from 75 to 90 cm (30 to 36 in) from the stem, and three to five of the strongest and best placed of these are selected and the rest cut off. The selected shoots are cut back to about two-thirds their length and this type of pruning is continued for the next few years with gradually reduced severity until the tree head is formed. Other forms in which fruit trees may be trained are standard, half-standard, espalier, cordon and fan.

Established mature trees may have the variety substituted by grafting. This is done either by framework grafting or by top-grafting methods which are described in Chapter 12.

Half-hardy fruits

This group comprises peaches and nectarines, vines and figs which, although usually grown under glass in this country, are sometimes cultivated quite successfully in the open.

PEACHES AND NECTARINES (*Prunus persica* varieties). Like the other tree fruits peaches are propagated by budding and grafting on to special rootstocks. The recommended stocks are Brompton and common mussel both of which are used for plums.

VINES (*Vitis vinifera*) Vines are usually propagated from 'eyes'. These consist of pieces of ripened stems of the current season's growth about 2.5 cm (1 in) long, each having a good bud at its centre. The 'eyes' are inserted horizontally with the bud uppermost in 7.6 cm (3 in) pots filled with compost such as JIP₁. 'Eyes' potted in January and kept in a warm greenhouse are ready for planting in borders in April. Usually, how-

ever, they are potted first into 11 cm (4½ in) pots and next into 15 cm (6 in) pots where they remain for the first year. In autumn the new growth is cut back to two buds. The second year the plants are kept growing in a warm greenhouse and are frequently sprayed with tepid water. They are finally potted into 20 or 25 cm (8½ or 10 in) pots. In the following autumn, increased ventilation is given and less water, to ripen off the canes which are supported with a bamboo. Two-year-old canes from pots are planted in early spring.

Vines may also be raised from hardwood stem cuttings planted in the open. The cuttings are secured during the winter when pruning, and are cut into lengths of 23 to 30 cm (9 to 12 in). These may be tied in bundles and heeled in to about half their depth in moist sand. In March trim off a fraction of an inch from the stem bases and plant the cuttings 8 to 10 cm (3 to 4 in) apart in trenches 18 cm (7 in) deep. Tread the soil in firmly. Growth usually starts in June and if 30 to 45 cm (12 to 18 in) of shoot growth is made the new plants may be transplanted to their permanent positions in autumn. If only a limited amount of growth is made then the young vines should be left undisturbed until the following autumn.

Some weak-growing varieties of vines succeed best when grafted on to stronger growing varieties. Thus, Muscat of Alexandria is sometimes used as a rootstock for Muscat Hamburgh or Madresfield Court. Whip-and-tongue grafting is the method used, and inarching, or grafting by approach, is also practised. Vines may also be increased by layering.

FIGS (*Ficus carica*) Figs are easily increased by cuttings, layers and suckers. Cuttings are most usual and consist of short-jointed, well-ripened shoots of the previous season's growth, cut into pieces about 15 cm (6 in) long, preferably with a heel. Insert the cuttings in pots containing light sandy compost in early spring and plunge them in peat in a warm greenhouse, preferably in a propagating frame provided with bottom heat. Cuttings root quickly and plants so raised may produce fruit before they are one year old. Young shoots may also be layered in a warm greenhouse where they quickly take root.

21
Raising Vegetable Crops and Culinary Herbs

Most vegetable crops are produced from seed but a few of the perennial kinds such as rhubarb and seakale are propagated by vegetative means.

Seed is usually sown in drills as this method facilitates weed control whether by hoeing or hand weeding. Chemical weed control is now used extensively in vegetable growing and is often very successful. Thus propachlor (Herbon Orange) the residual soil herbicide descheribed in Chapter 6 is effective for the control of weeds in seedling crops of brassicas as well as onions and leeks. Simazine controls weeds in asparagus, rhubarb, sweet corn and broad beans.

Another advance in the raising of vegetable crops is the use of pelleted seed as described in Chapter 6. This allows seed-spacing and obviates transplanting with many crops such as brassicas.

How the different kinds are increased

ARTICHOKE, GLOBE (*Cynara scolymus*) Usually increased from suckers which are found at the bases of established plants. These are simply cut off with a sharp knife and planted in the open 10 cm (4 in) deep and about 60 cm (2 ft) apart in spring.

ARTICHOKE, JERUSALEM (*Helianthus tuberosus*) These are very hardy and increase naturally by tubers. Plant from February to April in rows 75 cm (30 in) apart, allowing 30 cm (12 in) between the tubers.

ASPARAGUS (*Asparagus officinalis*) Asparagus is raised from seed, which may be sown in March or early April in drills 45 cm (18 in) apart. After covering with soil the bed should be firmed. The seed is slow to germinate. The seedlings are thin-ned out to 8 cm (3 in) apart, leaving the strongest. During the summer hoe frequently, and water in dry weather. The following spring the plants may be moved to their permanent site.

BASIL, SWEET (*Ocimum basilicum*) This is an annual herb which is raised by sowing seed in heat in April and planting out in early June.

BEAN, BROAD (*Vicia faba*) For early crops seed of the long pod varieties is sown in November, except in cold districts. A sowing may also be made in a warm border in January or February. Another method is to start the plants in a cold frame and transplant them later. Alternatively, early sowings may be protected by covering the rows with cloches for a few weeks. Broad beans are usually sown in double rows, the seed being spaced 23 cm (9 in) apart each way. The double rows should be 75 cm (2½ ft) apart.

BEAN, KIDNEY OR FRENCH (*Phaseolus vulgaris*) Sow in the open in late April or early May in drills 5 cm (2 in) deep and 45 cm (18 in) apart. Space the seed 8 cm (3 in) apart and thin out to 15 cm (6 in) apart. Early crops are secured by sowing in March in cold frames in drills across the frame 30 cm (1 ft) apart. When the plants are through, cover the frames with mats on frosty nights. Double rows sown in April may be given protection for a few weeks with cloches. Early crops of French beans may be had if a few seeds are sown in 20 cm (8 in) pots in a warm greenhouse in December, January and February.

BEAN, RUNNER (*Phaseolus multiflorus*) Sow in the open in early May. Earlier sowings may be made in greenhouses or frames for transplanting, or early sowings in rows in the open may be protected by cloches. For training on sticks or poles the seeds are spaced 12 cm (4½ in) apart in double rows 23 cm (9 in) apart, the double rows being not less than 1 m (3 ft) apart. The plants are usually thinned out to leave 23 cm (9 in) between them.

BEETROOT (*Beta vulgaris*) Sow in late April or early May in rows 30 cm (12 in) apart. Thin out the plants to about 8 cm (3 in) apart. Wider spacing usually results in poorer quality and lower yield.

BORAGE (*Borago officinalis*) This annual herb may be sown in March or April and thinned out to 30 to 38 cm (12 to 15 in) apart.

BORECOLE OR KALE (*Brassica oleracea acephala*) Sow thinly in early April in rows 30 cm (12 in) apart. Transplant when large enough, leaving 75 cm (2½ ft) between the rows and spacing the plants 60 cm (2 ft) apart.

BROCCOLI (*Brassica oleracea italica*) Sow according to variety from mid-April to mid-May in drills as advised for borecole. Plant out from June to mid-July. Space the plants about 68 cm (2 ft 3 in) apart each way.

BRUSSELS SPROUTS (*Brassica oleracea bullata gemmifera*) For early crops the seeds may be sown in a greenhouse or cold frame, the plants being set out in April or May. In some districts autumn sowings are made in sheltered borders for spring transplanting; seed for the maincrop is usually sown in drills in March. These are transplanted in June. Set the plants 75 cm (2½ ft) apart each way.

CABBAGE (*Brassica oleracea capitata*) Summer cabbage of the round Primo type is usually sown in cold frame in January or February. Further sowings may be made in the open in April for autumn produce. Winter cabbage such as January King and Savoys are sown in April or early May. Spring cabbage is sown the second or third week in July. Transplant, when large enough to handle, 45 cm (1½ ft) apart.

CARDOON (*Cynara cardunculus*) This vegetable is grown in trenches like celery. The seed is sown at 45 cm (18 in) stations in prepared patches in the trenches. Sow in late April and cover the seeds with a flower pot until they have germinated. Later, thin the seedlings to leave one at each station.

CARROT (*Daucus carota*) For early crops the seed may be sown broadcast thinly in a cold frame in October or January and left there until ready for pulling. The stump-rooted type may be sown in a warm border in March. Maincrop carrots are usually sown in April or early May. Make the drills 1 cm (½ in) deep and 30 cm (1 ft) apart. When sown evenly and not too thickly there is no need to thin out.

CAULIFLOWER (*Brassica oleracea botrytis cauliflora*) Early cauliflowers are usually sown in September in a cold frame. When large enough to handle they are pricked out in another frame and wintered there. Alternatively the plants may be potted into 8 cm (3 in) pots which are placed either in a cold frame or on a greenhouse floor and wintered there. These plants are set out in March or April. Another sowing may be made in a heated greenhouse in January or February in boxes. The seedlings are pricked out into other boxes, hardened off and planted in the open in April. Further sowings for autumn produce may be made outside in April.

CELERY (*Apium graveolens*) Sow in a heated greenhouse in February or March and prick out into boxes or a cold frame. Harden off and plant out in trenches (or in beds for self-blanching celery) in late May or early June. Seed may also be sown in a cold frame in March. Pelleted seed can be used, space-sown to avoid pricking out.

CELERIAC OR TURNIP-ROOTED CELERY (*Apium graveolens rapaceum*) Sow about mid-March and give the same treatment as advised for celery. Plant out on the flat 30 cm (12 in) square in early June.

CHICORY (*Cichorium intybus*) Sow in late May in rows 30 cm (1 ft) apart. Thin the plants out 15 to 23 cm (6 to 9 in) apart.

CHIVES (*Allium schoenoprasum*) Lift and divide the clumps every three or four years in the spring. May also be grown from seed.

ENDIVE (*Cichorium endivia*) This autumn and winter salad crop should be sown for succession from June until mid-August. Spacing is similar to lettuce.

HORSERADISH (*Cochlearia armoracia*) Easily propagated from root cuttings about 8 cm (3 in) long. These are inserted in holes made with a long dibber and afterwards filled in with soil. Plant in spring 45 by 30 cm (18 by 12 in) apart.

LEEK (*Allium porrum*) The seed is usually sown about mid-March in shallow drills 38 cm (15 in) apart. The seedlings are transplanted when large enough to handle 15 cm (6 in) apart in 10 cm (4 in) deep drills made about 30 cm (12 in) apart. To secure plants earlier a sowing may be made in a cold frame or greenhouse in January or February.

LETTUCE (*Lactuca sativa*) With the aid of glass protection and artificial heat it is possible to have lettuce for cutting all the year round. Suitable varieties must be used for the different seasons and conditions. Approximate dates of sowing the principal varieties are as follows:

1. Sow outdoors from early March until late July at two- to three-week intervals, using varieties such as Webb's Wonderful or Avondefiance.

2. Sow the varieties Imperial or Arctic King about mid-August in the open. Transplant in late September to a sheltered situation.

3. Sow the varieties Seaqueen or Emerald in a cold frame or greenhouse in early September and a further sowing of the same varieties in October in slight heat. Transplant the seedlings, when large enough to handle to a heated greenhouse border.

4. Sow the variety May Princess in a cold frame in Early October. Transplant the seedlings in December or January to a cold frame or greenhouse. Pelleted seed allows space sowing.

Lettuce is normally grown in rows from 23 to 30 cm (9 to 12 in) apart, with the same distance between the plants.

MARJORAM, POT (*Origanum onites*) A perennial herb which is increased by sowing seed in April. Plant about 23 cm (9 in) apart.

MARJORAM, SWEET (*Origanum majorana*) This herb is treated as an annual and is usually sown in heat in March or April. Transplant in May 23 cm (9 in) apart.

MINT (*Mentha* spp.) Easily propagated by division of the underground stems in spring or autumn. Summer cuttings also root in a shady situation in the open.

ONION (*Allium cepa*) Bulb onions are produced by sowing in the open in March. Earlier sowings may be made in heat and the plants set out when large enough to handle. They can also be sown in August and the seedlings transplanted the following spring. The variety White Lisbon is often sown in August for use as a salad in spring. Bulb onions are grown in rows 30 to 38 cm (12 to 15 in) apart, the plants being spaced about 10 cm (4 in) apart. Pelleted seed is available for space-sowing.

Onion sets (small bulbs) may be planted 10 cm (4 in) apart in rows 30 cm (12 in) apart in March and April for summer cropping. Sets can be produced by sowing onion seed quite thickly on a patch of poor ground in mid-May. The young plants are kept free of weeds but are not fed or otherwise encouraged to grow vigorously. They can be harvested in August, dried and replanted the following spring.

Potato onions and shallots (*Allium ascalonicum*) are increased by division of the bulbs which are lifted in the autumn and planted in spring 15 to 23 cm (6 to 9 in) apart.

PARSLEY (*Petroselinum crispum*) Sow in March for summer cutting, in June for winter produce, and again in August for use the following spring. The seed is slow to germinate. Drills should be spaced 30 cm (12 in) apart and the plants thinned to 8 to 10 cm (3 to 4 in) apart.

PARSNIP (*Pastinaca sativa*) Sow in February or March in drills 38 to 45 cm (15 to 18 in) apart. Thin out plants to 15 cm (6 in) apart.

PEA (*Pisum sativum*) In mild districts early peas may be sown in November. Normally successional sowings are made from early March until mid-June. Early sowings may be protected by cloches. Use early, mid-season and later varieties according to season, but early varieties should be chosen for late autumn sowing. Ordinary V-shaped drills about 6 to 8 cm (2½ to 3 in) deep, and spaced about the same distance apart as the plants are expected to grow, are suitable, but flat-bottomed drills may be used if preferred and the seeds sown in a 15 cm (6 in) wide band, each band containing two or three rows.

RADISH (*Raphanus sativus*) Successional sowings can be made from February to August in the open. Earlier sowings may be protected in cold frames or under cloches. Seed is usually sown broadcast or in drills 15 cm (6 in) apart.

RHUBARB (*Rheum rhaponticum*) Divide the old roots in autumn or spring. Each portion should have at least one crown or 'eye' and is called a set. The sets are usually planted in rows 1 m (3 ft) apart with 75 cm (2½ ft) being allowed between the sets. Rhubarb can also be raised from seed, but does not always come true to variety. Sow in March in drills 2.5 cm (1 in) deep and 30 cm (12 in) apart. Thin the plants to 23 cm (9 in) apart, and transplant in autumn or spring.

SAGE (*Salvia officinalis*) Easily increased from seed sown outdoors in April. Soft cuttings root readily in spring or early summer and plants so raised are often preferable to those from seed. Plant 38 cm (15 in) apart.

SALSIFY (*Tragopogon porrifolius*) Sow in drills 2.5 cm (1 in) deep and 30 cm (12 in) apart in April.

SCORZONERA (*Scorzonera hispanica*) Increased in the same manner as salsify.

SEAKALE (*Crambe maritima*) Usually increased from root cuttings or thongs about 5 mm ($\frac{1}{4}$ in) thick and 15 cm (6 in) long. These are prepared in autumn, tied in bundles and laid in sand. The sprouts which appear on each cutting in spring should be rubbed off to leave just the strongest. The thongs are then planted about 2.5 cm (1 in) below the surface in rows 45 cm (18 in) apart with 38 cm (15 in) between the cuttings. Seakale may also be raised from seed sown in the open in March.

SPINACH (*Spinacia oleracea*) A succession of summer spinach is maintained by sowing at two- to three-week intervals from early March until the end of June. Winter spinach is provided by making a sowing at about mid-August and another at mid-September. Sow in drills 2.5 cm (1 in) deep and 30 cm (12 in) apart, and thin to 15 cm (6 in).

SPINACH, NEW ZEALAND (*Tetragonia expansa*) Seed is sown under glass in late March and the seedlings are potted. Harden off and plant out in late May, the plants being 60 cm (2 ft) apart and the rows 1 m (3 ft) apart.

SPINACH BEET (*Beta vulgaris cicla*) Two sowings are made for succession, one in April and one in late July. Sow in drills 38 cm (15 in) apart and thin out the plants to 15 to 23 cm (6 to 9 in) apart.

THYME (*Thymus vulgaris*) The common thyme is easily raised from seed sown in April or by division in spring. Plant about 10 cm (4 in) apart.

TURNIP (*Brassica rapa*) Turnips and swedes are raised by sowing seed from March until May. Earlier sowings may be made in frames or protected by cloches. The plants are usually thinned out to about 10 cm (4 in) apart and are sown in rows 45 cm (18 in) apart.

VEGETABLE MARROW (*Cucurbita pepo*) Sow seed singly in 8 cm (3 in) pots in late March or early April. Germinate and grow in heat for a period. Harden off and plant out towards the end of May. The seed may also be sown in a cold frame in April or in the open in May.

Half-hardy vegetables

TOMATO (*Lycopersicon esculentum*) This is one of the most valuable crops grown in greenhouses. The propagation stage is most important and to make a bad start with tomatoes will invariably affect the final results. For this purpose select a position in a greenhouse where light conditions are at a maximum. Choose an F_1 hybrid variety and use John Innes or a soilless seed compost. The seeds should be spaced about 1 cm ($\frac{1}{2}$ in) apart in trays or pots and covered with about 3 mm ($\frac{1}{8}$ in) of compost. The containers should be placed preferably on open slatted benches.

During the night a temperature of 18°C (65°F) should be maintained, which may rise to over 21°C (70°F) during the day. When the cotyledons have expanded, which is usually 8 to 10 days after sowing, set the seedlings singly into 10 cm (4 in) plastic pots using JIP compost or a soilless equivalent. From now on the night temperature is lowered to 16°C (60°F) but is allowed to rise to over 21°C (70°F) during sunny days. Watering should be carried out frequently and in sunny weather especially may be necessary two or three times a day. During dull weather, however, over-watering must be avoided.

When the seedlings are well established in the pots they should be fed with a proprietary liquid fertilizer specially recommended for this purpose. Space out the plants as they grow so that the leaves do not overlap. Planting out in borders or into larger pots is recommended when about 50% of the plants are showing their first truss flowers.

Artificial illumination is beneficial and is described in Chapter 7.

CUCUMBER (*Cucumis sativus*) Cucumbers are another useful crop whose propagation is somewhat less exacting than tomatoes. The best method is to sow the seed singly in 8 cm (3 in) pots. A suitable compost consists of equal parts good loam and well-rotted stable manure to which is added the JI base fertilizer at the same rate as used in the JIP compost.

Fill each pot to about three-quarters its capacity and press a plump seed into the centre. Cover with about 1 cm ($\frac{1}{2}$ in) compost, give a good watering and then place glass and paper over the pots.

Maintain a temperature of about 21°C (70°F) and a moist atmosphere, and germination occurs in 36 to 48 hours. The glass and paper are then removed, but avoid exposing the seedlings to strong sunlight. Continue to grow in a temperature of 18 to 21°C (65 to 70°F) and water when necessary.

When the roots have reached the sides of the pots, pot on into 15 cm (6 in) pots using a similar compost with a treble dose of fertilizers. The plants should then be staked with 60 cm (2 ft) canes. Plant out in the beds before there is any risk of a check to growth in the pots. Cucumber seedlings may be artificially illuminated for a period of about three weeks with considerable advantage.

Unlike tomatoes there is no need to allow a daily period of six hours' darkness. Illumination may be continuous without any harmful effect and results in earlier crops.

MELON (*Cucumis melo*) Melons may be raised in exactly the same manner as cucumbers except that the composts used need not be so rich. JIS or a soilless seed compost is suitable for seed sowing and the seedlings may be potted on in JIP compost or a soilless equivalent. Artificial illumination is beneficial.

22
How Decorative and House Plants are Increased

Under this heading a wide range of ornamental foliage and flowering plants is included. Most of them are increased by seed or cuttings involving the general principles outlined in previous chapters. Thus seeds are usually sown in JIS compost and potted into JIP compost with a single, double or treble dose of fertilizers according to the type of plant and the length of time it is likely to remain undisturbed. Soilless composts made specifically either for seed sowing or for potting may also be used.

The seeds of greenhouse plants are usually germinated in a temperature of 13 to 18°C (55 to 65°F), and a similar temperature is allowed for the propagation of cuttings.

How the different types are increased

ABUTILON (Indian mallow) Half-ripe cuttings taken in spring or autumn will root with bottom heat. Seed germinated in a temperature of 18 to 21°C (65 to 70°F) is another method.

ACACIA (wattle, mimosa) Insert half-ripe cuttings with a heel in July in pots containing a peaty compost. Plunge the pots in a propagator; when rooted pot up singly and grow in cool conditions. Seeds sown when ripe in a mixture of peat and sand germinate freely in a temperature of 16°C (60°F).

ACALYPHA (copper leaf) Cuttings should be rooted in brisk heat in spring.

ACHIMENES Sow seed in early spring with care, as it is very small. Germinate in a temperature of 16 to 18°C (60 to 65°F). Prick out the seedlings in light peaty compost and grow in a similar temperature. Cuttings secured from plants started in

heat should be inserted in a propagator in a compost of peat and sand with bottom heat. Leaf cuttings also root under similar conditions when their stalks are inserted in the compost. Another method is to rub the scales off the corms and sow them in the same compost, covering lightly. Given bottom heat the scales should soon start to grow.

AECHMEA Readily increased from suckers which develop naturally. These should be potted and kept in a warm, humid atmosphere until rooted.

AEONIUM Cuttings in heat strike readily at any time.

AGAPANTHUS (African lily) Divide the plants just as growth begins in the spring and pot the rooted portions.

AGAVE (American aloe) Offsets which are freely produced are a simple means of increase.

AGLAONEMA Propagated by division or cuttings in warm, humid conditions.

ALLAMANDA Cuttings of previous season's growth can be rooted in a propagator in spring.

ALOCASIA The rhizomes can be divided in spring.

ALOE Increase is similar to Agave.

ALPINIA (Indian shell flower) Divide in spring when growth is starting.

ANANAS (pineapple) Remove and pot up suckers in spring or summer. The tufted growths from the top of the fruits can be removed at the same time and rooted in a warm propagator.

ANTHURIUM Divide the roots in March when repotting. Sow seed in a mixture of chopped sphagnum moss, charcoal and sand, temperature 21 to 27°C (70 to 80°F).

APHELANDRA Increase from nodal cuttings of

semi-mature wood or soft shoots with a heel in bottom heat of 21°C (70°F).

ARALIA Soft cuttings made from sideshoots can be rooted in a propagator. Root cuttings taken in spring will also grow. These are made about 5 cm (2 in) long and are inserted in pots of sandy soil. The pots are placed in a propagator; bottom heat promotes rooting.

ARAUCARIA (Norfolk Island pine) Seed of *A. heterophylla* (syn. *A. excelsa*) can be sown in heat in spring; cuttings of shoot tips will root in a propagator in summer (see p. 106).

ARDISIA (spear flower) Seed may be extracted from ripe berries and sown immediately in heat. Cuttings taken between March and September will strike in heat.

ASCLEPIAS (milkweed) Divide the roots in October or April. Take cuttings in spring and root in a propagator. Sow seed in gentle heat in February and treat as an annual. Pot the seedlings and cuttings when they are ready.

ASPARAGUS The well-known asparagus ferns *A. plumosus* and *A. sprengeri* are raised from seed. Sow in spring in JIS compost. Pot singly and grow in a fairly moist, warm atmosphere. Roots may also be divided when repotting large plants in spring.

ASPIDISTRA Divide and repot during the growing season. Shade for few days afterwards.

BEGONIA Most of the begonias may be increased from seed, which is sown in early spring. The seed is very small and requires a fine surface and no covering. Germinate in a temperature of 16 to 18°C (60 to 65°F). The ornamental-leaved kinds such as *B. rex* are increased by leaf cuttings as described in Chapter 11. The fibrous-rooted kinds, like *B. socotrana* and its varieties, such as John Heal and Gloire de Lorraine, are readily increased from stem cuttings secured from plants started in heat in early spring. The propagation of tuberous-rooted begonias is given in Chapter 18.

BELOPERONE (shrimp plant) Easily increased from cuttings in moderate bottom heat.

BERTOLONIA Readily increased from seed or cuttings in heat. Leaves pegged down on sandy compost with cuts across the veins will also produce new plants in heat.

BIGNONIA (cross vine) These climbing plants are increased by soft cuttings in spring inserted in a propagator. Young shoots may also be layered in late summer.

BILLBERGIA Increased by division, the suckers being twisted off the main stem. Pot separately and keep in brisk heat for a week or two.

BORONIA Insert cuttings of firm young shoots in a propagator in August, preferably with bottom heat.

BOUGAINVILLEA Cuttings taken with a heel in March or April strike in a propagator.

BOUVARDIA Increase is by soft cuttings about 5 cm (2 in) long taken in March and inserted in pots of sandy compost. Plunge the pots in a propagator with bottom heat. Root cuttings about 2.5 cm (1 in) long will grow when planted in pans in spring. Divide the plants when repotting.

BROWALLIA Raised from seed sown in spring or in July for late winter and spring flowering.

BRUNFELSIA Cuttings root readily in moderate heat. Seed should be germinated in a temperature of at least 21°C (70°F).

CACTI AND SUCCULENT PLANTS This large and variable group of interesting plants is not difficult to propagate. Seed and cuttings are the principal methods. Seeds may be sown at any time if heat is available, but spring sowing is generally preferable. Sow thinly in well-drained pans of fine compost and cover according to the seed size. Plunge the pans in peat or ashes and cover them with paper to prevent rapid drying. Maintain a temperature of 10 to 13°C (50 to 55°F) and remove the paper immediately germination is observed. Prick out the seedlings in light compost as soon as they are large enough to handle.

Generally speaking, cacti and succulents are very easy to root from cuttings, but some species, such as the spherical kinds, do not provide any material for this purpose. Any branching plant, however, will usually provide cuttings, and offsets can be secured from others. Usually the cuttings should be dried for a few days before insertion. A compost of peat and sand is quite suitable, but pure sand also serves the purpose. There is no need to provide humid conditions, but slight shade may be beneficial. Cuttings of strong-growing species such as the opuntias may be inserted singly in pots. Many kinds can be raised from leaf cuttings. Grafting is sometimes used for a few special kinds.

CALADIUM Propagated by division of the tubers in early spring.

CALATHEA (zebra plant) Easily increased by division.

CALCEOLARIA (slipper flower) The shrubby species are increased by cuttings which are taken in autumn and inserted in a cold frame or in boxes kept in a cool greenhouse. Herbaceous calceolarias are raised from seed sown in July. Prick out and pot on as required.

CALLISTEMON (bottle-brush tree) Take cuttings of firm young shoots in summer and insert in pots containing a compost of peat and sand. Place the pots in a propagator.

CAMPANULA (bellflower) C. isophylla is popular for hanging baskets and should be raised annually from cuttings taken in summer.

CANNA (Indian shot) Sow seed singly 2.5 to 5 cm (1 to 2 in) deep in small pots after soaking in water for 24 hours. Sow in February or March, and keep the pots in a temperature of 21°C (70°F). Named varieties are raised by dividing the roots at potting time.

CARNATION Perpetual-flowering carnations are propagated by cuttings which should be selected with care. The best are those made from sideshoots found at about the centre of the flowering stems. Cut below a node and remove the lower leaves. Insert the cuttings in sharp, clean sand and maintain a temperature of 10 to 13°C (50 to 55°F). Bottom heat is also beneficial. Water after insertion, and subsequently keep the medium nicely moist. Keep the cuttings warm at first, but free ventilation should be given when rooting begins in a few weeks time. Rooted cuttings are potted into 8 cm (3 in) pots using a gritty compost. Carnation cuttings of this type are normally taken from December to March, but they may also be rooted at other seasons.

Malmaison carnations are best increased by layering in June. The rooted layers are potted in August.

CASSIA (senna plant) Cuttings of half-ripe wood will strike in a propagator. Seed is also used.

CELOSIA (cockscomb) This annual makes an attractive pot plant. For this purpose the seed is sown in a warm greenhouse in March and is first pricked out and then potted on as necessary.

CEROPEGIA Easily increased by cuttings or by seed if obtainable.

CHLOROPHYTUM (spider plant) Readily increased by division or plantlets.

CHORIZEMA These evergreen Australian shrubs are increased by sowing seed in a warm greenhouse in March or by heel cuttings taken in July and rooted in a propagator.

CHRYSANTHEMUM The marguerite, C. frutescens, is increased from cuttings removed from old

plants in April and rooted in a propagator. See also Chapter 16.

CINERARIA Raised from seed, which is sown for succession from April until June. JIP compost is suitable for seed sowing. Pot the seedlings into JIP₂ and later into the same compost in larger pots.

CISSUS Propagated by softwood cuttings about 5 cm (2 in) long with a heel. Bottom heat is an advantage.

CITRUS Cuttings can be rooted in a propagator in summer; seeds can be sown in spring in a temperature of 16°C (60°F).

CLIVIA Sow seeds in a warm greenhouse in spring and keep the seedlings growing until they flower. Division of old plants can be carried out when repotting in February.

COBAEA (cup and saucer vine) C. scandens is easily raised from seed sown in a warm greenhouse in early spring. The seedlings flower the same year. The variegated type is raised from cuttings of firm sideshoots taken July or August and inserted in a propagator.

CODIAEUM (croton or Joseph's coat) Easily increased by stem cuttings, which are inserted singly into pots. The latter are plunged in peat in a warm propagating frame with brisk bottom heat.

COLEUS (flame nettle) Plants of variable types may be raised from seed sown in gentle heat in early spring. Cuttings of young shoots can be rooted at any time in sandy compost and kept warm. Cuttings are essential to increase special forms.

COLUMNEA Cuttings from firm shoots will strike in a warm propagating frame.

CORDYLINE The greenhouse species are increased by cutting the main stem into pieces 2.5 to 5 cm (1 to 2 in) long. These are inserted in a propagator in sandy soil. The half-hardy species such as C. australis are easily raised from seed.

CRASSULA Sow seed in a warm house in early spring and keep the seedlings rather dry. Stem or leaf cuttings strike readily in pots of sandy compost stood on the greenhouse staging.

CRINUM Increase by seed or offsets. The large seeds should be sown singly in pots and germinated in a temperature of 18 to 21°C (65 to 70°F).

CROSSANDRA Easily increased by cuttings given bottom heat throughout the year.

CRYPTANTHUS Offsets should be separated and potted up singly in spring in brisk heat.

161

CUPHEA (Mexican cigar flower) Increased from seed sown in March in gentle heat. Cuttings of young shoots can be struck in spring and summer.

CYCLAMEN Propagated by sowing seed in August or January. Sow seeds 3 mm ($\frac{1}{4}$ in) deep and 2.5 cm (1 in) apart in pans or trays or individually in Jiffy 7's. Allow a temperature of 10 to 13°C (50 to 55°F). Shade the seedlings from bright sun and pot when large enough to handle.

CYPERUS (umbrella plant) Propagated by seed or division in moderate heat.

CYTISUS (florists' genista) Usually increased by cuttings of sideshoots taken in spring with a slight heel and inserted in pots, which are placed in a propagator. Seed may be sown in spring.

DIEFFENBACHIA (dumb cane) Increased by basal suckers or by short stem pieces like cordylines in brisk heat.

DIONAEA (Venus's fly trap) Large plants can be divided in spring; seeds can be sown on a mixture of peat and sphagnum moss in spring – germinate at a temperature of 18°C (65°F).

DIZYGOTHECA (false aralia) Stem cuttings can be rooted in a propagator in summer; root cuttings will shoot if planted in a propagator in spring.

DRACAENA (dragon plant) Propagated by stem pieces as for cordylines; also by 'toes' – large pieces of root cut into 2.5 cm (1 in) lengths and dibbed vertically into a moist mixture of peat and sand. Maintain a temperature of 18° C (65° F).

EPACRIS (Australian heath) Take cuttings from shoot tips and insert in pots of sandy, peaty compost in August. Place the pots in a warm propagator. Seeds are sown, when ripe, in a similar compost and germinated at 10 to 13°C (50 to 55°F).

EPIPHYLLUM Sow seeds in sandy compost and germinate in a temperature of 18°C (65°F) in spring; cuttings can be rooted in pots of sandy compost in spring and summer.

ERANTHEMUM Take cuttings from March to June and strike them in propagator with bottom heat.

ERICA (heath) Tip cuttings 2.5 cm (1 in) long are inserted in pots in sandy peat in spring. Keep warm and use gentle heat. May be rooted under mist.

EUPHORBIA (poinsettia) Cuttings are taken in early spring and struck in a propagator with brisk heat.

EXACUM Sow seeds in a temperature of 16°C (60°F) in spring for summer and autumn flowering; in summer for winter flowering.

FATSHEDERA Cuttings can be rooted in a propagator in summer.

FATSIA (false castor oil) F. japonica can be raised from stem cuttings about 5 cm (2 in) long in spring. Insert in a propagating frame. Seeds sown singly in pots germinate at 18°C (65°F).

FERNS Many species can be raised from spores as described in Chapter 7. Division when repotting, usually in spring, is generally adopted. The well-known Asplenium bulbiferum is increased by minute plantlets which develop on mature fronds. The fronds may be pegged down on light compost like a leaf cutting, or the plantlets may be taken off singly and carefully inserted in 7 cm ($2\frac{1}{2}$ in) pots.

FICUS (rubber plant) F. elastica is propagated by short nodal stem pieces each with a leaf which are inserted in pots or boxes stood in a warm propagating frame or by air-layering. Other species are increased by stem cuttings rooted in a propagator in spring and summer.

FITTONIA Divide plants in spring; cuttings can be rooted in a propagator in spring or summer.

FRANCOA (bridal wreath) Divide the plants when repotting. Summer cuttings will strike in a propagator. Seeds may be sown in pots of sandy compost in March or April.

FREESIA Separate the offsets when repotting in the autumn. Usually six to eight corms are inserted in a 12 cm (5 in) pot. Easily raised from seed sown when ripe or in April. Seedlings usually flower before they are one year old.

FUCHSIA Usually raised from cuttings of soft young shoots. These are taken in spring or in autumn and inserted in pots of sandy compost. Seed may be sown in spring in pans and germinated in a temperature of about 16°C (60°F).

GARDENIA These beautiful evergreen shrubs are increased by cuttings. Secure young shoots in January and insert singly in pots of sandy compost. Place the pots in a propagator. Plants so raised will flower the following winter, but later-rooted cuttings provide a succession of bloom.

GERBERA (Transvaal daisy) Usually increased by seed in gentle heat, but for special strains cuttings consisting of sideshoots with a heel will strike in a propagator with bottom heat.

GESNERIA Can be increased naturally by division of the tubers. Cuttings made from young basal shoots will root in early spring in brisk bottom heat. Seed is another method but requires care as it is extremely fine. Sow as for begonias.

GLORIOSA Remove offsets from established plants in spring. Seeds can be sown individually in small pots and germinated in a temperature of 24°C (75°F) in spring.

GLOXINIA (sinningia) The gloxinia may be increased by seed which is sown in pans of light compost in March. Basal cuttings are easily rooted in pots of sandy compost placed in a warm frame. Leaf cuttings provide another method.

GREVILLEA (silk oak) *G. robusta* is increased from seed sown in February in sandy compost in a temperature of about 21°C (70°F). Insert cuttings of young shoots taken with a heel in spring in pots of sandy compost. Place the pots in a warm propagator.

GYNURA Cuttings can be rooted in a propagator in spring.

HELIOTROPIUM (cherry pie, heliotrope) Cuttings of soft young shoots are inserted in spring around the edges of well-drained pots of light compost. Place in a propagator until rooted and then pot singly. Pinch the young plants two or three times to induce a bushy habit. Cuttings to be trained as standards are not stopped until they reach the required height.

HIBISCUS (rose mallow) Cuttings from mature current season's growth will strike in bottom heat in the autumn. Other methods are autumn or spring layering and grafting on to seedling stocks. Hybrids may be raised from seeds sown in spring in a temperature of 18°C (65°F).

HIPPEASTRUM Secure offsets when repotting old bulbs in early spring. Seed should be sown in heat in March and potted on when necessary. Seedlings do not reach flowering age until they are three years old.

HOWEIA Sow seeds in pots placed in a propagator in spring. Temperature: 24°C (75°F).

HOYA (wax flower) Cuttings of *H. carnosa* made from shoots of the previous year's growth are inserted in pots containing a compost of peat and sand in the spring. The pots are placed in a frame kept at a temperature of 21 to 23°C (70 to 75°F). Young shoots may be layered during the summer by pegging them into pots of peat and sand.

HUMEA Sow seeds in pots or pans in July and place in a cold frame or greenhouse. Pot when large enough to handle.

HYDRANGEA *H. macrophylla* var. *hortensia* (syn. *H. hortensia*) the common hydrangea, is increased by cuttings. Young shoots are taken in spring and

inserted singly in small pots of sandy compost. Place in a warm frame until rooted and afterwards grow in a cool greenhouse or frame. Pot as required.

HYPOCYRTA (clog plant) Cuttings of firm young shoots can be rooted in a propagator in summer.

HYPOESTES (polka dot plant) Seeds of *H. sanguinolenta* can be sown in a temperature of 16° C (60° F) in spring.

IMPATIENS (balsam) Greenhouse species are easily raised from seed but special strains or varieties are readily propagated from softwood cuttings.

IPOMOEA (morning glory) Annual species are raised from seed sown in March in a warm greenhouse. Perennial species are increased from cuttings made from sideshoots during the summer. Root in a propagator. *I. batatas* (the sweet potato) is increased by division of the tubers in February, or from cuttings.

IRESINE Firm young shoots can be taken as cuttings in spring and rooted in a propagator.

JACARANDA (green ebony tree) Cuttings in a sand and peat mixture with bottom heat root readily in early spring. Seed when available will germinate in a temperature of 21 to 23°C (70 to 75°F).

JACOBINIA Cuttings can be rooted in a propagator in spring or summer.

JASMINUM (jasmine) Greenhouse species are propagated from soft young cuttings taken with a heel in spring. These are inserted in a warm frame.

KALANCHOË Seed may be raised in pots or pans of light compost in a temperature of 16 to 18°C (60 to 65°F). The seedlings require care in handling. Cuttings are made from young sideshoots after flowering. Both these and leaf cuttings strike readily in a sandy soil.

KENNEDYA (coral creeper) Seed sown in spring will germinate in a temperature of 13 to 16°C (55 to 60°F). Cuttings of young sideshoots will strike in pots of sandy compost placed in a warm frame.

LACHENALIA (Cape cowslip) Bulbs increase naturally and are usually repotted in August and planted 2.5 to 5 cm (1 to 2 in) apart. Seeds may also be sown in the spring in heat and are not difficult to germinate. Leaf cuttings are also successful.

LANTANA Propagated by seed sown in spring in a temperature of 21 to 23°C (70 to 75°F). Cuttings of young sideshoots inserted round the sides of a pot in spring root in a propagator. Half-ripe cuttings taken in August or September and treated similarly will also root.

LAPAGERIA Can be raised from seed sown in pans and kept in a warm greenhouse. Young shoots can also be layered indoors in spring or summer.

MARANTA Divide established plants in spring.

MAURANDIA Usually raised from seed and treated as an annual. Cuttings, however, will strike in a propagator in autumn and the plants kept in a greenhouse over winter.

MEDINILLA Cuttings of firm young growths can be rooted in a propagator in spring.

MONSTERA (shingle plant, Swiss cheese plant) Cut the stem into pieces three joints long and root these in a warm propagating frame.

MUSA (banana) Seeds can be sown in a temperature of 24°C (75°F) in spring after first being soaked in water for 24 hours. Suckers can be removed from mature plants in spring and potted up individually.

NEOREGELIA Offsets can be removed in spring and summer and rooted in a propagator.

NERINE (Guernsey lily) Offsets are secured when repotting, usually in August or September, and are potted in 8 to 15 cm (3 to 6 in) pots.

NERIUM (oleander) Cuttings made from firm wood are potted singly in sandy compost in spring or summer. Keep in a propagating frame until rooted.

ORCHIDS These comprise a large and varied group of plants, and propagation must be related to the usual cultural practice for each kind. Many orchids can be increased by division, this being done when repotting, which is normally undertaken just when growth is starting in spring. Genera treated in this way include cattleyas, calanthes, cymbidiums, cypripediums, dendrobiums and masdevallias. Propagation by seed is very difficult and undertaken by specialists only.

PACHYSTACHYS Cuttings of firm shoots can be rooted in a propagator in summer.

PALMS These plants are usually increased by seed which is sown in spring or when available. Sow in well-drained pans or pots, which should be plunged in peat and kept in a propagating frame with a temperature of 21 to 23°C (70 to 75°F). A few species can be propagated from suckers, which are potted in spring.

PANCRATIUM Propagation is by potting offsets which are secured when repotting old bulbs in March.

PANDANUS (screw pine) Remove suckers in spring and keep warm until the roots are established.

PASSIFLORA (passion flower) Cuttings 10 to 15 cm (4 to 6 in) long with a heel are made from young shoots in spring. These are inserted in a warm propagating frame. Seed is another method.

PELARGONIUM (geranium) The different types are raised from cuttings which are cut off below a joint and are made 10 to 15 cm (4 to 6 in) long. Dib five cuttings around the edge of an 11 cm (4½ in) pot containing sandy compost and place on the greenhouse staging. Under closer, more humid conditions damping-off is probable. Seed of hybrids can be sown in January and February in a temperature of 21°C (70°F).

PEPEROMIA Cuttings of short stem pieces with a single joint will root in pots stood on greenhouse staging. Under closer, more humid conditions, damping-off is probable. Seed is another method.

PHILODENDRON Increase by cuttings as for monstera.

PILEA Cuttings of firm young shoots can be rooted in a propagator in summer.

PIMELEA Make cuttings from young shoots in spring and insert round the side of a pot filled with a compost of peat and sand. Keep in a propagating frame until rooted.

PLECTRANTHUS Divide large plants in spring; take cuttings in summer and root in a propagator.

PLUMBAGO (Cape leadwort) Easily propagated by stem or root cuttings with bottom heat.

PLUMERIA (frangipani) Cuttings of firm shoots can be rooted in a propagator in summer

POLIANTHES (tuberose) Remove and pot up offsets in spring.

PRIMULA Seed is the normal method of increasing greenhouse primulas and is usually sown in April or May for winter flowering. *P. malacoides*, however, is sometimes sown in July to provide flowers in spring. Sow the seeds in pots and prick out the seedlings when they are large enough to handle. Later they are potted into 8 cm (3 in) pots and finally into the 12 cm (5 in) size. A temperature of 13 to 16°C (55 to 60°F) is suitable for seed germination and 10 to 13°C (50 to 55°F) for growing the seedlings. Double-flowered primulas are raised from cuttings or division.

PROTEA Sow seeds in peaty compost in a cool greenhouse in spring. Soak in hot water for one hour before sowing.

REINWARDTIA Cuttings of firm shoots can be rooted in a propagator in spring.

RHOEO Cuttings of young shoots can be rooted in a propagator in spring or summer.

RHODANTHE These 'everlasting' annuals are sometimes grown in pots. The seed may be sown in September or March and germinated in a warm greenhouse or frame.

RHOICISSUS Propagate by cuttings in the spring, as for cissus.

SAINTPAULIA (African violet) Mixed strains of *S. ionantha* are raised from seed, which is usually sown in early spring. Sow in pots and keep them in a temperature of 18 to 21°C (65 to 70°F). Prick out and pot as the seedlings grow. Shading is important. Leaf cuttings provide another method of increase which should be used on all desirable individual types.

SALVIA *S. splendens* is usually raised from seed which is sown in the spring in pots or boxes. Germinate in a temperature of 16 to 18°C (60 to 65°F) and pot singly in 8 cm (3 in) pots when the seedlings are large enough to handle. Most salvias may be increased from cuttings. These are secured from plants of the previous year retained over winter in mild heat and started into growth in spring. Insert the cuttings in pots which are placed in a propagating frame.

SANSEVIERIA Easily increased by suckers. Also by leaf cuttings, the long leaves being cut into lengths of about 8 cm (3 in) which are inserted in sandy compost in heat. The yellow-edged form of *S. trifasciata* known as *S. laurentii* must be propagated by division if its yellow stripes are to be retained (Plate 8).

SAXIFRAGA (mother of thousands) *S. sarmentosa* increases naturally by creeping stems or runners. From these the plantlets are removed and inserted in small pots.

SCHIZANTHUS (butterfly flower, poor man's orchid) For summer flowering sow the seeds in early spring in a warm greenhouse. August is the time to sow when the plants are required to bloom in spring. Prick the seedlings out and pot up as necessary.

SCINDAPSUS (devil's ivy) Cuttings of firm shoots can be rooted in a propagator in summer; suitable stems can be layered at the same time.

SELAGINELLA (creeping moss) Cuttings inserted in pots root readily in a propagating frame. When rooted several cuttings may be potted together to form a single specimen. Division when repotting is another method of increase.

SETCREASEA Cuttings can be rooted in a propagator in summer.

SMITHIANTHA (temple bells) Divide rhizomes in spring and replant individually; sow seeds in late spring in a temperature of 18°C (65°F).

SOLANUM *S. capsicastrum*, the popular winter cherry, is usually increased from seed which is sown in February in heat using JIP or a soilless equivalent. Prick out the seedlings and pot them up when they are large enough, first into 8 cm (3 in) pots and finally into the 8 cm (5 in) size. Pinch the plants to induce a bushy habit. The winter cherry may also be increased by cuttings inserted in pots placed in a propagating frame in March. The climbing species such as *S. jasminoides* are increased from soft young cuttings in spring struck in pots in a propagator.

SOLEIROLIA (baby's tears, mind-your-own-business) Divide in spring or summer.

SPARMANNIA (African hemp) This is increased by cuttings secured from plants cut back in early spring. Root in a propagator in a temperature of 16 to 18°C (60 to 65°F).

SPATHIPHYLLUM Divide mature plants in spring; sow seeds in spring in a temperature of 24°C (75°F).

STEPHANOTIS Cuttings of firm shoots can be rooted in a propagator in summer.

STRELITZIA (bird of paradise flower) May be raised from seed which, when sown, should be kept in a high temperature preferably with bottom heat. Old plants may be divided when being repotted in spring.

STREPTOCARPUS (Cape primrose) Normally raised from seed which is sown in pans in January or February for winter flowering. Seed sown in July will flower the following summer. Germinate in a temperature of 16 to 18°C (60 to 65°F) and prick out and pot as necessary. Special forms of Streptocarpus may be increased by dividing the old plants. Leaf cuttings will grow if inserted in a sandy compost.

STREPTOSOLEN *S. jamesonii* is increased by cuttings made from soft young shoots in spring or summer and inserted in a propagator. Rooted cuttings are potted singly and pinched once or twice to encourage a bushy habit.

STROBILANTHES Cuttings inserted in a warm propagating frame root readily.

SWAINSONA (Darling river pea) Cuttings of young shoots root readily in summer in a propagating frame. Seeds germinate quickly when sown in March in a temperature of 16 to 18°C (60 to 65°F).

THUNBERGIA Propagate by seed sown in heat or by half-ripe cuttings in a warm frame.

TIBOUCHINA Cuttings can be rooted in a propagator in summer.

TOLMIEA (pick-a-back plant) Divide large plants in spring; remove leaves with plantlets attached and root in a propagator in summer.

TRACHELIUM (throat-wort) *T. caeruleum* is raised from seed sown in spring or July in a warm greenhouse. Pot the seedlings up singly when they are large enough and pinch the shoots several times to induce bushy growth.

TRADESCANTIA (spiderwort, wandering jew) Easily increased by cuttings of young shoots inserted in a propagating frame.

TORENIA Usually raised from seed sown in March or April and treated as a greenhouse annual. It may, however, be increased by cuttings in a propagator.

VALLOTA (Scarborough lily) Repotting should be carried out in autumn when necessary. Offsets may then be removed and are potted singly in small pots.

YUCCA Species such as *Y. aloifolia* can be increased by offsets. Also 8 cm (3 in) lengths of stem inserted in small pots and placed in a propagator will give rise to roots and shoots.

ZANTEDESCHIA (arum lily) Repotting and division are usually carried out in August or September. Suckers which are readily produced are potted singly and started in mild heat. Seed provides another method and may be sown in spring in a warm greenhouse.

ZEBRINA Easily increased from cuttings in moderately warm conditions at any time.

23
Plant Pests and Diseases in Relation to Propagation

The usual methods of controlling plant pests and diseases apply to a large extent in propagation. It is particularly important, however, to prevent damage at the early and vulnerable stages of a plant's life. Moreover, there are certain pests and diseases which confine their attacks mainly to seeds or young plants, and are therefore of special interest to the propagator.

It should be unnecessary to emphasize that good cultivation, proper manuring, crop rotation, the destruction of weeds and general garden hygiene play a large part in checking the depredations of pests and damage caused by diseases.

In the open a well-prepared seedbed which has been adequately fertilized encourages rapid and healthy growth of seedlings so that the phase of their greatest susceptibility to injury from pests and disease is shortened. The same is true under glass when a good compost is used for seed growing.

Carelessness in sowing may cause overcrowding of the young plants and increased susceptibility to disease. Faulty covering may result in some seeds being buried so deeply that they give rise to weak unhealthy seedlings if they germinate at all.

Crop rotation helps in preventing the accumulation of certain pests and diseases which attack particular crops. Examples are the potato root eelworm (*Heterodera rostochiensis*) and white root rot of onions (*Sclerotium cepivorum*).

Both in the open and under glass, weeds, should be destroyed as early as possible, for not only do weeds check the growth of crop plants but they also act as hosts to certain pests and diseases.

In relation to the control of pests and diseases, garden hygiene is of major importance. Rubbish of all kinds may give shelter to various pests and act as a starting point to disease. Decaying vegetable matter and such-like material should be kept well away from cultivated plants. In greenhouses, particularly, the strictest hygiene is essential. Refuse of any sort should never be allowed to accumulate under the staging or in odd corners.

A greenhouse or frame used in propagation should have at least one good annual cleaning and disinfecting. This involves scrubbing down the interior of these structures and washing the glass clean both inside and out. The interior may be scrubbed down with a good disinfectant such as diluted Jeyes' Fluid. All containers used in propagation should also be washed and sterilized, as well as canes and other equipment. The sterilization of soil intended for propagating purposes reduces the risk of disease. Soil sterilization is described in Chapter 7.

As a general rule, greenhouses occupied by plants should be fumigated periodically to keep down general pests like aphids and leaf hoppers. For this purpose a gamma-HCH preparation can be recommended. This is available in the form of smoke canisters or pellets which, on being ignited, discharge smoke impregnated with the insecticide into the air.

Seed treatment
This method is used extensively in the control of pests and diseases. In some cases the seedsman will carry out the treatment. This applies for instance to what is called the thiram soak treat-

ment which is used for certain seedborne fungal diseases including broccoli stem canker (*Phoma lingam*) and celery blight (*Septoria apiicola*). Some seedsmen will also undertake the fumigation of onion and leek seed to ensure freedom from seed-borne eelworm infection while other firms will dress onion seed with a chemical called dicloran which helps in preventing white rot (*Sclerotium cepivorum*).

Seed dressings which can be applied by the gardener include orthocide for peas, broad beans and sweet corn to prevent seed rot due to soil fungi. Brassica seed may be dressed with gamma-HCH against flea beetles and cabbage root fly, while the fungicides thiram or orthocide may be added to these dressings to protect the seed against soil-borne fungi.

There are still a number of diseases which can be seed-transmitted, and against these seed-dressings are ineffective. Among these, virus diseases are prominent, examples being lettuce and tobacco mosaic virus which affects tomatoes. Most seedsmen can supply mosaic-tested lettuce seed and such varieties are usually specified in catalogues. With tomatoes it is possible to secure anti-virus tested seed at an extra charge. Both these diseases cause severe deterioration of the affected crops, but unfortunately there are other sources of infection as well as seed.

Fungus and bacterial diseases which can be seed-transmitted include halo blight of French beans (*Pseudomonas phaseolicola*) which can cause severe damage to the crop; black ring spot of brassicas (*Mycosphaerella brassicicola*) and grey mould of lettuce (*Botrytis cinerea*). Such diseases cannot be prevented by seed treatment. All this indicates how important it is to secure seed only from reliable seedsmen who will always avoid as far as possible saving seed from diseased plants.

Seedling enemies

APHIDS Greenfly will attack plants of any age. They spread virus diseases as well as reducing vigour and secreting sticky honeydew. Spray with pirimicarb or derris.

FLEA BEETLES There are several species of these which attack cruciferous seedlings, particularly brassicas. Injury occurs as a pitting of the seed-leaves which, if severe, may cause the seedling to die. A seed dressing as mentioned above is usually an adequate control, otherwise dusting the seed-bed with gamma-HCH is very effective.

PEA AND BEAN WEEVIL (*Sitona lineata*) Damage to seedling broad beans and peas is due to the feeding of the beetles on the leaf margins which take on a scalloped appearance. If damage is significant, dust with gamma-HCH.

CABBAGE AND TURNIP GALL WEEVIL (*Ceutorrhynchus pleurostigma*) This pest causes rounded galls on the roots of young brassica plants—not to be confused with club root. The best control is to destroy affected plants when transplanting and to avoid growing canbrassicas too frequently on the same soil.

CUTWORM (*Agrotis* spp. etc.) These attack a wide range of seedling crops usually in the open. The dull greyish brown typical caterpillars bite at the base of seedlings so that they may be cut right through. Where the pest is suspected treat the soil surface with bromophos when setting out susceptible plants such as brassicas. Keep down weeds which may act as alternate hosts, and cultivate the soil to expose the pest to the birds.

LEATHERJACKET (*Tipula paludosa*) These well-known soil pests are the larvae of the crane fly or daddy-long-legs. They feed on the roots of almost any plant and are particularly injurious to seedlings. Fork gamma-HCH into the soil at planting time and observe hygiene points noted for cutworms.

MILLIPEDES There are several species which attack the roots of seedlings and also burrow into seeds like peas, beans and beetroot. They cause damage both outside and under glass. A method of control for outdoor crops is to dust with gamma-HCH along open seed drills after sowing. Under glass normal soil sterilization will effect control, but if a growing crop is attacked water the roots with a diluted solution of gamma-HCH. This solution must not be allowed to come into contact with foliage.

SLUGS AND SNAILS Several species are injurious to seedlings. Control measures include the use of proprietary pelleted baits containing metaldehyde or methiocarb; these are fairly effective but under wet conditions the latter is preferable. Lay the baits under pieces of slate or tile. Alternatively trap the pests on a small scale by using upturned grapefruit skins, or yoghurt cartons half filled with beer and sunk into the ground around susceptible crops.

WIREWORMS (*Agriotes* spp.) The well-known larvae of the click beetle attack both seed and seedlings of many plants. Under glass wireworms bore into the stems of young plants such as lettuce and tomatoes, while outside innumerable seeds and plants are attacked.

Control measures include hygiene because, as the eggs are usually laid in weedy land in the autumn, clean cultivation renders infestation less probable. Gamma-HCH seed dressings are recommended below for other pests and also deter wireworm. Another control is to work into the soil about 15 g per sq m ($\frac{1}{2}$ oz per sq yd) gamma-HCH dust. Soil treatment with this insecticide, however, involves the risk of tainting certain root crops such as carrots and potatoes if these are grown up to 18 months after treatment. Root fly Three important pests of seedlings come under this heading. They are onion fly (*Delia antiqua*), cabbage root fly (*Delia brassicae*) and carrot fly (*Psila rosae*). All of these may be partially or sometimes completely controlled by seed dressings as already mentioned, but further treatment may be necessary and is essential at any rate after transplanting brassicas. Onion fly can be prevented by dusting along the seedling rows with calomel. When transplanting brassicas the soil may be dusted with bromophos. Felt or plastic discs placed around the plants may stop the pest from laying its eggs. To prevent carrot fly dust the seed with gamma-HCH or dust along the rows fortnightly with the same substance from mid-May onwards. Avoid thinning the seedlings as this can encourage attacks. A paraffin-soaked rag dragged over the plants at intervals reputedly discourages attack.

DAMPING-OFF DISEASE OF SEEDLINGS (*Pythium sp.* and *Rhizoctonia solani*) These are common diseases of seedlings grown under glass in humid conditions. Fortunately they are not nearly so severe as they once were. This is because seed is now usually sown in soil sterilized by steam or chemicals, or in soilless composts where the risk of infection is greatly reduced.

Even when the soil is sterilized, water can be a source of infection and for seedlings it is usually best to use water direct from the mains. If tanks are used they must be kept scrupulously clean and chemically sterilized periodically. Should the disease appear, keep the seedlings on the dry side and avoid a close, humid atmosphere by judicious ventilation. Watering with diluted Cheshunt compound is a preventive measure. Various types of damping-off sometimes attack plants in greenhouse beds, in frames or in the open.

Plant health in relation to vegetative propagation

Many diseases and pests may be present in propagating material such as tubers, corms, bulbs, runners and cuttings. In these circumstances the disease or pest is very likely to be inherited by the new plants. Of course, there are several diseases and pests which often attack this type of propagating material after it has been planted.

Virus diseases A wide range of plants is affected by different virus diseases which are often spread naturally through the agency of sucking insects, such as aphids and thrips, as they move from diseased plants on to healthy ones. Virus diseases are usually, therefore, less severe in cold, wet districts where there is a low insect population. This is the reason why Scottish and Irish 'seed' potatoes give better results than 'seed' produced in southern England.

General symptoms of virus diseases on plant foliage are spots, concentric rings, mottling and blotching in colours of yellow, white or light green. Occasionally the shape of the leaf alters as in 'reversion' of blackcurrants, and sometimes there is a leaf crinkling or rolling. Invariably viruses cause plant deterioration, with a loss of vigour and decreased yield. Affected strawberries, for instance, become severely dwarfed.

Common virus diseases are blackcurrant reversion, raspberry mosaic, strawberry leaf-crinkle, strawberry leaf-roll, potato mosaic, tomato mosaic and dahlia mosaic. Spotted wilt is a disease which attacks a wide range of plants, including tomatoes, winter cherry, arum lilies, begonias, chrysanthemums, dahlias and delphiniums.

Propagation can be a means of spreading virus diseases and it is therefore most important to propagate susceptible plants from stocks known to be healthy. Special methods of doing so have been described in the case of raspberries and strawberries. Sometimes it is inadvisable to use home-produced material for propagation. This usually applies in the case of potatoes. Dahlias suffer severely from virus diseases and, to keep a stock healthy, plants for propagation should be selected annually. This consists in marking those which are strong and vigorous and have a normal leaf appearance during the growing season. When a private gardener is in doubt regarding the health of any plant he should seek expert advice before attempting propagation.

When purchasing plants of strawberries, raspberries and blackcurrants the amateur should always ask for certified plants. This means that such plants have been certified by the Ministry of Agriculture as being true to variety and that they conform to a certain standard of health.

It is interesting to note that a few virus diseases may be seed-transmitted. Examples are lettuce mosaic, and common bean mosaic, which attacks dwarf beans.

Fungus and bacterial diseases A number of diseases may be transmitted from the old to the new plants by diseased material. An example is mint affected with the rust *Puccinia menthae*. Affected plants can, however, be rendered free from disease by washing the soil from the roots and immersing the plants in water heated to 37.5°C (110°F) for 10 minutes. Afterwards, plunge the treated plants in cold water and plant them in sterilized soil or in soil which has not been used previously for growing mint.

LEAFY GALL (*Corynebacterium fascians*) is a bacterial disease that attacks many plants including carnations, chrysanthemums, dahlias, gladioli, pelargoniums and verbascums. This disease causes growth to assume a dense mass of small shoots which fail to develop. Infected plants should be destroyed to prevent their being used in propagation.

CROWN GALL (*Agrobacterium tumefasciens*) is a similar type of disease to leafy gall. It causes galls of various sizes to develop on plant stems or roots, usually near soil level. About seventy different species are known to be susceptible, including fruit trees and bushes and dahlias, but the disease is not regarded as serious. Obviously, however, affected material should not be used for propagation.

Bulbs, corms and tubers are liable to be affected by a considerable number of diseases and can thus transmit such infection in propagation. Gladioli are subject to the disease dry rot (*Sclerotinia gladioli*), hard rot (*Septoria gladioli*) and neck rot (*Bacterium marginatum*). In storage under damp conditions the corms may be affected with green mould (*Penicillium gladioli*) and core rot (*Botrytis gladiorum*). Narcissus bulbs are liable to infection by bulb rot or basal rot (*Fusarium oxysporum*), storage rot (*Penicillium narcissi*), white mould (*Ramularia vallisumbrosae*) and smoulder (*Botrytis narcissicoli*). Fungi which affect tulips are grey bulb rot (*Sclerotium tuliparum*), tulip fire (*Botrytis tulipae*) and other diseases which cause decay of the bulbs. Similarly, potato tubers are liable to various infections such as dry rot (*Fusarium caeruleum*) and blight (*Phytophthora infestans*). Most other bulbs, corms and tubers are affected with similar diseases.

Insects and related pests Perhaps the worst pests which are likely to be transmitted in propagation are eelworms. These are microscopic worms which live and feed in plant tissues. The chrysanthemum eelworm (*Aphelenchoides ritzemabosi*) affects the leaves, causing them to turn brown and fall off, the lower ones first.

The stem and bulb eelworm (*Ditylenchus dipsaci*) attacks a wide range of plants, but those which suffer most are narcissus bulbs, which decay when badly affected; herbaceous phlox, whose stems become twisted and distorted, and strawberries. The last-named are also affected by strawberry eelworm (*Aphelenchoides fragariae*) which causes leaf puckering of leaves and poor growth.

Warm-water treatment is widely used to control eelworm. Thus, chrysanthemum stools after flowering are washed free of soil and are then steeped in water at 40°C (115°F) for five minutes. Narcissus bulbs are immersed in warm water at 37.5°C (110°F) for three hours, and strawberry runners are given 20 minutes at the same temperature. This treatment will also destroy the tarsonemid mite.

After treatment the plant material should be plunged in cold water and the chrysanthemums and strawberries replanted in sterilized soil or soil known to be free from eelworm. In this way healthy propagating material is ensured. Phlox affected with eelworm should be propagated from root cuttings, as the eelworms live in the stems but do not enter the roots.

Narcissus flies (*Merodon equestris* and *Eumerus tuberculatus*) are other pests which attack this genus. The flies lay their eggs near the plants in summer and the maggots which hatch out enter the developing bulbs, where they remain until after the crop is lifted. These maggots can be killed in the bulbs by the warm-water treatment used for eelworms. If bulb fly maggots alone are present, one hour's immersion is sufficient.

Index

Abelia, 71, 105
Abies, 14, 105
Abutilon, 105, 159
Acacia, 105, 159
    false, 12
Acaena, 129
Acalypha, 159
Acantholimon, 103, 129
Acanthus, 122
Acer, 105
    budding, 93
    layering, 104
    mist propagation, 71
    seed stratification, 39
Achillea, 122, 129
Achimenes, 159
Acidanthera, 136
Aconite, winter, 137
Aconitum, 60, 122
Acorus, 128
Actinidia, 105
Adenophora, 122
Adonis, 122, 129
    aestivalis, 142
    vernalis, 39
Aechmea, 159
Aeonium, 159
Aesculus, 105
Aethionema, 129
Agapanthus, 122, 159
Agave, 159
Ageratum houstonianum, 143
Aglaonema, 159
Agriotes, 168–9
Agrostemma 'Milas', 142
Agrotis, 166
Ailanthus, 39, 82, 105
Ajuga, 129
Akebia, 105
Albizia, 105
Alchemilla, 122
Alder, 105–6
    white, 108
Alisma, 128
Allamanda, 159
Allium, 136, 155, 156
Almond, 39
Alnus, 105–6
Alocasia, 159
Aloe, 159

American, 159
Alonsonoa, 143
Alpines, 129–35
    cuttings, 75, 77, 82
    division, 60
    layering, 103
    seed saving, 34
    transplanting, 83–84
Alpinia, 159
Alstroemeria, 34, 122
Althaea, 143
Alum-root, 124
Alyssum, 129, 142
Amaranthus caudatus, 143
Amaryllis belladonna, 136
Amelanchier, 91, 106
Ampelopsis, 106
Anagallis, 129
Ananas, 159
Anaphalis, 129
Anchusa, 82, 122, 142–3
Andromeda, 106
Androsace, 34, 129
Anemone, 122, 136
Angelica tree, 106
Annuals, 13
    half-hardy, 142–3
    hardy, 141–2
Anthemis, 122, 129
Anther, 30
Anthericum, 34, 122, 136
Antholyza, 136
Anthurium, 159
Anthyllis, 106, 129
Antirrhinum, 28, 75, 129, 143
Aphelandra, 159–60
Aphids, 168
Aponogeton, 128
Apple, 112
    budding, 93
    cuttings, 74, 80
    grafting, 85, 86, 89, 91
    rooting, 68
    rootstocks, 85, 149
    seed stratification, 39
    vegetative increase, 17
Apricot, 39
Aquilegia, 122, 129
Arabis, 60, 130
Aralia, 106, 160

false, 162
Araucaria, 14, 106, 160
Arbutus, 71, 106
Arctostaphylos, 106
Arctotis grandis, 143
Ardisia, 160
Arenaria, 130
Arisaema, 136
Arisarum, 122
Aristolochia, 106
Armeria, 130
Arnebia, 130
Aronia, 106
Arrowhead, 128
Artemisia, 122
Artichoke
    globe, 154
    Jerusalem, 62–63, 154
Arum, 122
    bog, 128
Aruncus, 128
Arundinaria, 106
Asclepias, 122, 160
Ash, 110
Asparagus, 38, 154, 160
Asperula, 130, 142
Asphodeline, 122
Aspidistra, 160
Asplenium bulbiferum, 162
Aster, 71, 122, 127, 130
    China, 143
    Mexican, 142
Astilbe, 123, 130
Astragalus, 130
Astrantia, 123
Atriplex, 106, 142
    hortensis, 142
Aubrieta, 60, 71, 130
Aucuba, 71, 74, 106
Auricula, 130
Auxins, 67–68
Avens, 123, 131
    mountain, 130
Azalea, 39, 71, 77, 114
    evergreen, 80
Azara, 106

Babiana, 136
Baboon root, 136
Baby blue-eyes, 142

Baby's tears, 165
Bacteria, 170
Ballota, 106
Balsam, 35, 143, 163
Bamboo, 106
Banana, 164
Baptisia, 123
Barrenwort, 123
Bartonia, 142
Basil, sweet, 154
Bay, sweet, 111
Bean, 37
  bog, 128
  broad, 154, 168
    diseases, 168
    seed, 15, 32, 35
  French, 154, 168
  kidney, 154
  runner, 154–5
  tree, Indian, 107
Bearberry, 106
Beard tongue, 132
Bear's breech, 122
Bee balm, 124
Beech, 110
  southern, 112
Beetroot, 155
  germination, 37
  pests, 168
  seed, 33, 44
Begonia, 82, 83, 136, 160
  cuttings, 14
  germination, 39–40
  sowing, 56
  tubers, 62–63
Bellflower, 123, 130, 161
  Chinese, 125
Bellis perennis, 39, 130
Bellium, 130
Bells of Ireland, 142
Bellwort, 123, 125, 130
Beloperone, 160
Benches, 21, 70, 72
Berberidopsis, 106
Berberis, 34, 61, 80, 106
Bergenia, 123
Bertolonia, 160
Betula, 106
Biennials, 13, 15, 143–5
Bignonia, 160
Billbergia, 160
Birch, 106
Bird of paradise flower, 165
Birds, 19
Blackberries, 148
  cuttings, 81, 83
  layering, 100
  mist propagation, 12
  natural propagation, 12
Blackcurrants, 146–7
  cuttings, 74
  disease, 169, 170
  layering, 103
  rooting, 14
  stock plants, 84
Bladder nut, 117
Bladder senna, 85, 108

Blanket flower, 123, 142
Bleeding heart, 123
Blight, 168, 170
Bloodroot, 125
Bluebell, 137, 140
Bluets, 131
Boilers, 21
Borage, 42, 155
Borecole, 155
Boronia, 160
Botrytis, 168, 170
Bottle-brush tree, 107, 161
Bougainvillea, 160
Bouvardia, 160
Box, 107
Brachycome iberidifolia, 143
Brassica, 32, 33, 51, 155, 168
Bridal wreath, 110, 162
Broccoli, 155, 168
Brodiaea, 136
Brompton rootstock, 150
Broom, 34, 80, 84, 109
  butcher's, 114
  hedgehog, 131
  Spanish, 117
Browallia, 143, 160
Brunfelsia, 160
Brunnera, 123
Brussels sprouts, 155
Bryophyllum, 82
Buckthorn, sea, 95, 111
Budding, 93–96
  fruit rootstocks, 150
Buddleia, 71, 106
Bugbane, 123
Bugle, 129
Bugloss, 122
Bulbils, 64
Bulbocodium, 136
Bulbs, 34, 63–65, 136–40
Bulrush, false, 128
Buphthalmum, 123
Bupleurum, 107
Burdock, *11*
Butomus, 128
Buttercup, 12, 128, 139
Butterfly flower, 165
Buxus, 71, 107

Cabbage, 155
  disease, 168, 169
  germination, 38
  sowing under glass, 56
Cactus, 12, 160
  cuttings, 76, 77
Caladium, 160
Calamint, 130
Calamintha, 130
Calathea, 160
Calceolaria, 75, 130, 160–1
Calendula officinalis, 141, 142
Calla, 128
Callicarpa, 107
Callistemon, 107, 161
Callistephus chinensis, 143
Calluna, 71, 107
Callus, 66–67

Calochortus, 136
Caltha, 128
Camassia, 136
Cambium, 13; 66, 67, 86
Camellia, 71, 83, 107
Campanula, 123, 130, 143, 161
  cuttings, 82
  seed, 34
Campion, 124, 131, 142
Campsis, 107
Candytuft, 131, 141, 142
Cane, dumb, 162
  Whangee, 113
Canker, 168
Canna, 38, 161
Canterbury bells, 34, 143
Caragana, 107
Cardamine, 130
Cardiocrinum, 136
Cardoon, 155
Carlina, 130
Carnation, 123
  cuttings, 14, 77
  layering, 103
  Malmaison, 161
  perpetual, 161
  seed, 35
Carpels, 31
Carpenteria californica, 107
Carpinus, 107
Carrot, 155
  germination, 37
  pests, 169
  seed, 33, 34
Caryopteris, 107
Cassia, 161
Cassinia, 107
Cassiope, 107
Castanea, 107
Castor oil, false, 110, 162
Catalpa, 107
Catananche, 123
Catchfly, 135
Catmint, 124
Cat-o'-nine tails, 128
Cauliflower, 58, 155
Ceanothus, 80, 107
Cedar, 107
  Japanese, 109
Cedrus, 107
Celastrus, 107
Celeriac, 155
Celery, 155, 168
  turnip-rooted, 155
Celmisia, 130
Celosia 143, 161
Celsia cretica, 145
Centaurea, 123, 141, 142
Centaurium, 130, 142
Ceutorrhynchus pleurostigma, 168
Centranthus, 123
Cephalaria, 123
Cerastium, 130
Ceratostigma, 107
Cercis, 107
Geropegia, 161
Cestrum, 107

Chaenomeles, 107
Chalk plant, 124, 131
Chamaecyparis, 14, 107
    cuttings, 74, 82
Chamomile, 122, 129
Cheiranthus, 130, 135, 143–5
Chelone, 123
Cherry, 150
    flowering, 81, 93, 114
    Japanese, 17
    sweet, 114
    winter, 165
Cherry pie, 163
Chestnut, Spanish, 107
Chicory, 155
Chimonanthus praecox, 108
Chinese lantern, 125
Chionanthus, 108
Chionodoxa, 136
Chlorophytum, 12, 161
Choisya ternata, 71, 108
Chokeberry, 106
Chorizema, 161
Chrysanthemum, 125–7
    annual, 141
    coronarium, 142
    cuttings, 75, 77, 78, 125–6
    early flowering, 126–7
    frutescens, 161
    maximum, 60, 123
    morifolium, 125–7
    pests, 170
    rooting, 67
    segetum, 142
Cigar flower, Mexican, 162
Cimicifuga, 123
Cineraria, 161
Cinquefoil, 125, 132
Cissus, 161
Cistus, 71, 77, 108
Citrus, 161
Cladanthus arabicus, 142
Clarkia, 34, 141–2
Clematis, 92, 108
    cuttings, 73, 75
    grafting, 92
    mist propagation, 71
Cleome spinosa, 143
Clerodendrum, 108
Clethra, 108
Clianthus, 85, 108
Clivia, 136, 161
Cloches, 23
Clog plant, 163
Clones, 17
Clover, 135
    calvary, 142
Cobaea scandens, 161
Cockscomb, 143, 161
Codiaeum, 161
Codonopsis, 123, 130
Colchicum, 136–7
Coleus, 161
Colletia, 108
Collinsia bicolor, 142
Colt rootstock, 150
Columbine, 122, 129

Columnea, 61
Colutea arborescens, 85, 108
Comfrey, 125
Common mussel rootstock, 149–50
Composts
    blocking, 53
    cuttings, 78
    potting, 51–52
    seed, 51–52
    soilless, 55–56
Cone flower, 125
Conifers
    cuttings, 73, 80, 81
    grafting, 89
    juveniles, 14
    layering, 100
    mist propagatioon, 71, 72
    seed stratification, 39
Convallaria, 123
Convolvulus, 108, 130, 142
Copper leaf, 159
Coral berry, 106
Coral creeper, 163
Cordyline, 161
Coreopsis, 34, 123, 142
Corms, 63–65, 136–40
Corncockle, 142
Cornflower, 123, 141, 142
Cornus, 104, 108
Corokia, 108
Coronilla, 108, 130
Cortaderia, 123
Cortusa, 130
Corydalis, 123
Corylopsis, 108
Corylus, 108
Cosmos bipinnatus, 142
Cotinus, 101, 108
Cotoneaster, 108
    cuttings, 80
    mist propagation, 71
    seed, 34, 39
Cotula, 130
Cotyledon, 130
Cowslip, American, 131
    Cape, 163
Crab, wild, 112
Cranesbill, 123, 131
Crassula, 161
Crataegus monogyna, 39, 85, 108
Crepis, 123
Cress, rock, 130
    stone, 129
Crinum, 137, 161
Crocosmia, 123, 137
Crocus, 63, 137
    Chilean, 140
Crossandra, 161
Cross-pollination, 29, 31
Croton, 67, 161
Crown vetch, 108, 130
Cryptanthus, 161
Cryptomeria, 109
Cuckoo flower, 130
Cucumber, 33–34, 47, 157–8
Cunninghamia, 109
Cup flower, 132

Cuphea, 127, 162
Cupid's dart, 123
× Cupressocyparis, 109
Cupressus, 109
Currants, 74, 147
    flowering, 35, 74, 100, 114
Cuttings
    hardwood, 67, 73–74
    internodal, 75
    leaf, 82–83
    leaf-bud, 83–84
    root, 80–82
    root promotion, 66–68
    semi-hardwood, 80
    softwood, 74–80
    stem, 14, 66–80
    transplanting, 83–84
Cutworm, 168
Cyananthus, 34, 130
Cyclamen, 130, 137, 162
Cydonia, 85, 107
Cynoglossum amabile, 145
Cyperus, 128, 162
Cypress, 109
    false, 107
    Lawson's, 107–8
    Leyland, 109
    summer, 143
    swamp, 117
Cypripedium, 130
Cytisus, 71, 109, 162

Daboecia, 109
Daffodil, 139
Dahlia, 127
    cuttings, 75, 77, 127
    disease, 169
    division, 127
    rooting, 62, 67
    seed, 34, 35, 127
Daisy, 130
    globe, 131
    kingfisher, 142
    Livingstone, 143
    Michaelmas, 60, 122
    Swan river, 143
    Transvaal, 162
Damping-off, 169
Dandelion, 11
Daphne, 39, 71, 82, 100, 109
Dazomet, 53, 80
Deblossoming, 65
Decodon, 128
Delphinium, 39, 123, 141, 142
    cuttings, 75
    division, 60
    mist propagation, 71
    seed, 34
Desfontainea spinosa, 109
Deutzia, 71, 103, 109
Dianthus, 60, 71, 130, 142, 145
Dicentra, 123
Dichlobenil, 48
Dicloran, 168
Dicotyledons, 12–13
Dictamnus, 123
Dieffenbachia, 162

Dierama, 123
Diervilla, 109
Digitalis, 123, 145
Dimorphotheca, 127, 142
Dionaea, 162
Disbudding, 92–93
Disease, 167–70
    transmission, 16–18, 168, 170
Division, 12, 60–1, 62, 152
Dizygotheca, 162
Dogwood, 74, 108
Doronicum, 60, 123
Draba, 130
Dracaena, 13, 162
Dracocephalum, 130
Dragon plant, 162
Dragonhead, false, 125
Drills, 44–45
Drimys, 109
Dropping, 61, 103
Dryas, 130
Dutchman's pipe, 106

Eccremocarpus scaber, 109
Echeveria, 130–1
Echinops, 123
Echium, 143
Edelweiss, 131
Edraianthus, 131
Eelworm, 167, 170
Egg plant, 35
Elaeagnus, 71, 109
Elder, 117
Electric heaters, 21
Elephant's ears, 123
Elm, 93, 118
    witch, 118
Embothrium, 109
Endymion, 137
Enkianthus, 109
Epacris, 162
Epilobium, 40, 123, 131
Epimedium, 123
Epiphyllum, 162
Eranthemum, 162
Eranthis, 137
Eremurus, 123
Erica, 61, 109–10, 162
Erigeron, 42, 60, 123, 131
Erinacea, 131
Erinus alpinus, 129, 131
Eriobotrya japonica, 83
Eriogonum, 131
Eriophorum, 128
Eriophyllum, 123, 131
Erodium, 82, 131
Eryngium, 82
Erythrina, 110
Erythronium, 137
Escallonia, 71, 80, 110
Eschscholzia californica, 142
Etiolation, 101
Eucalyptus, 110
Eucomis, 137
Eucharis, 137
Eucryphia, 110
Euonymus, 39, 110

Eupatorium, 123
Euphorbia, 123, 145, 162
Euryops, 110
Everlasting flower, 131
Evodia, 110
Exacum, 162
Exochorda, 110

Fabiana, 110
Fagus sylvatica, 110
Fatshedera, 162
Fatsia, 110, 162
Felicia, 127–8, 142
Ferns, 18, 58–9, 162
Fertilizers, 42–3
Ficus carica, 153
    elastica, 115, 162
        cuttings, 83
        layering, 104, 116
Fig, 153
Figwort, cape, 113
Filberts, 62
Fir, silver, 105
Firethorn, 114
Fittonia, 162
Flag, 124, 128, 131
    sweet, 128
Flax, 131, 142
    New Zealand, 113
Flea beetle, 168
Fleabane, 123, 124, 131
Flowers, doubling of, 17
Fly pests, 168–70
Forget-me-not, 132, 145
    Chatham Island, 124
    rock, 132
    water, 128
Forsythia, 71, 74, 110
Fothergilla, 110
Foxglove, 123, 143
Frames, 22–3
    for cuttings, 77–78, 79–80
    for seeds, 58
Francoa, 110, 162
Frangipani, 164
Frankenia, 131
Fraxinus excelsior, 110
Freesia, 162
Fremontodendron, 110
Fringe tree, 108
Fritillary, 137
Fruit, 146–53
    budding, 150
    grafting, 150–3
    rootstocks, 149–50
    seed, 35
    trees, 148–53
Fuchsia, 110, 162
    Californian, 135
    cuttings, 75
    flower structure, 30
    mist propagation, 71
    stock plants, 84
Fumitory, 123
Fungus, 170

Gaillardia, 34, 123, 142

Gall, 170
Galanthus, 137
Galega, 123
Galtonia, 137
Gamma-HCH, 168
Gardenia, 14, 162
Garrya, 100, 110
Gas heaters, 21
Gaultheria, 80, 103, 110
Gazania splendens, 143
Genista, 39, 110, 162
Gentian, 34, 39
Gentiana, 131
Geranium, 123, 131, 164
    alpine, 82, 131
    cuttings, 76, 78
Gerbera, 162
Germander, 117, 135
Germination, 35–41
Gesneria, 162
Geum, 123, 131
Gilia tricolor, 142
Ginkgo biloba, 110
Gladiolus, 137–8
    bulbs, 63, 64
    de-blossoming, 65
    disease, 170
    seed, 34
Glaucium, 123–4
Globe flower, 125
Globularia, 131
Gloriosa, 163
Glory flower, Chilean, 109
Glory-of-the-snow, 136
Glory pea, 85, 108
Gloxinia, 51, 56, 163
Glyceria, 128
Glyphosate, 48
Goat's beard, 123, 130
Godetia 141, 142
Golden bells, 110
    chain, 111
    club, 128
    drop, 132
    rod, 125
Gooseberries, 147
    cuttings, 74, 75
    layering, 100, 103
    seed, 35
    stock plants, 84
Gorse, 84, 118
    Spanish, 110
Gourds, 35
Grafting, 85–93
    bench, 91–92
    framework, 86, 89, 96
    fruit rootstocks, 150–3
    inarching, 92–93
    inlaying, 88–89
    inverted L, 90, 91
    natural, 12
    rind, 87, 88
    saddle, 88
    side, 89, 90
    splice, 86, 87
    stub, 89, 90
    top, 89

tying and sealing, 91
under glass, 92
veneer, 89
wax, 91
wedge, 88
whip, 86, 87
whip and tongue, 86–87
woody plants, 14
Grass,
cotton, 128
manna, 128
pampas, 123
seed, 44
whitlow, 130
umbrella, 128
Green ebony tree, 163
Greenhouses, 20–25
for cuttings, 76–77, 79
for grafting, 92–93
for seeds, 51–59
fumigation, 51
sterilization, 167
Grevillea robusta, 163
Griselinia, 74, 110
Growing rooms, 24
Gum tree, 110
Gunnera, 128
Gynura, 163
Gypsophila, 92, 124, 131, 141–2

Haberlea, 82, 131
Habranthus, 138
Haemanthus, 138
Halesia carolina, 39, 110
Hamamelis, 110
Hardening-off, 15
Hawk's beard, 123
Hawkweed, 124
Hawthorn, 108
budding, 93
grafting, 85, 86
water, 128
Hazel, 39, 108
witch, 110
Heartsease, 135
Heated bins, 74
Heath, 103, 109, 162
Australian, 162
Irish, 109
Portuguese, 110
sea, 131
tree, 109
Hebe, 71, 74, 110
Hedera, 39, 110
Helenium, 124
Helianthemum, 71, 131
Helianthus, 60, 124, 142, 154
Helichrysum, 131, 142
Heliopsis, 124
Heliotrope, 75, 163
Heliotropium, 163
Helipterum, 142
Helleborus, 128
Hemerocallis, 34, 124
Hemlock, 118
Hemp, African, 165
agrimony, 123

Hepatica, 124
Herbicides, 48–9, 154
Herbon Orange, 48, 154
Herbs, 154–8
Heron's bill, 131
Hesperis matronalis, 145
Heterodera rostochiensis, 167
Heuchera, 24
Hibiscus, 104, 110–11, 163
Hickories, 14
Hieracium, 124, 131
Hippeastrum, 163
Hippophae, 95, 111
Hoes, 47–48
Hoheria, 111
Holly, 69, 111
sea, 123
Hollyhock, 143
Honesty, 145
Honeysuckle, 101, 109, 112
Hop, 142
Hormones, 67–68
Hornbeam, 107
Horse-chestnut, 93, 105
Horseradish, 155
Hosta, 124
Hottonia, 128
House plants, 159–66
Houseleek, 61, 135
Houstonia, 131
Houttuynia, 128
Howeia, 163
Hoya carnosa, 83, 163
Humea, 163
Humulus japonicus, 142
Hutchinsia, 131
Hyacinth, 64, 65, 138
Californian, 136
grape, 139
Hyacinthus, 138
Hybridization, 16, 27–31
Hybrids, 16, 27, 29, 30
graft, 85
seed, 15–16
Hydrangea, 75, 101, 111
climbing, 117
cuttings, 80
stock plants, 84
Hypericum, 40, 111, 128, 131
rock, 34, 131
Hypocyrta, 163
Hypoestes sanguinolenta, 163

Iberis, 129, 131, 141, 142
Ilex, 111
Immortelle flower, 142
Impatiens, 143, 163
Inarching, 92–3
Incarvillea, 124
Indian shot, 161
Indolyl acetic acid, 68
Indolyl butyric acid, 68
Inlaying, 88–89
Inula, 124
Ioxynil, 48
Ipheion, 138
Ipomoea, 142, 143, 163

Iresine, 163
Iris, 124, 128, 131
bulbs, 64
de-blossoming, 65
German, 60, 62
pseudacorus, 39
seed, 34
sibirica, 34
Ivy, 12, 110
devil's, 165
Ixia, 138

Jacaranda, 163
Jacobinia, 163
Jacob's ladder, 125, 132
Jasmine, 101, 163
bastard, 107
Chinese, 118
Jasminum, 80, 111, 163
Jeffersonia, 131
John Innes composts, 52
Joseph's coat, 161
Judas tree, 107
Juglans, 111
Juncus, 128
Juniperus, 14, 40, 111
Juveniles, 14

Kalanchoe, 163
Kale, 155
Kalmia, 111
Kansas gayfeather, 124
Kennedya, 163
Kerria, 17, 61, 111
Kirengeshoma, 124
Kniphofia, 124
Knives, 25
Knotweed, 132
Kochia scoparia, 143
Kolkwitzia amabilis, 111

+ *Laburnocytisus adamii*, 85
Laburnum, 34, 86, 91, 111
Lachenalia, 83, 163
Ladybell, 122
Lady's fingers, 106
Lady's mantle, 122
Lady's slipper, 130
Lagerstroemia, 111
Lamium, 124
Lantana, 163
Lapageria, 101, 163–4
Larch, 111
Larkspur, 34, 141
Larix, 111
Lathyrus odoratus, 142
Laurel, cherry, 114
common, 74
Portugal, 74, 114
spotted, 106
spurge, 109
Laurus nobilis, 111
Lauristinus, 118
Lavandula, 111
Lavatera, 111
trimestris, 142
Lavender, 75, 111

cotton, 117
Layering, 12, 99–104, 116, 133–4
   air, 104, 116, 121
   Chinese, 104
   continuous, 101
   herbaceous, 103
   mound, 101–3
   natural, 12
   serpentine, 101
   stool, 101–3
   tip, 100–1
Layia elegans, 142
Leadwort, 107
   Cape, 164
Leatherjacket, 168
Ledum, 103, 111
Leek, 155
   pests, 168
   seed sowing, 44, 51, 58
Leiophyllum, 111
Leontopodium alpinum, 131
Leopard's bane, 123
Leptospermum, 71, 111
Lettuce, 156
   disease, 68
   germination, 38, 40
   light requirements, 24
   seed, 35, 51, 56
Leucojum, 138
Leucothoë, 100, 112
Lewisia, 34, 35, 131
Leycesteria, 112
Liatris, 124
Ligularia, 124
Ligustrum, 112
Lilac, 117
   cuttings, 80
   grafting, 86
   layering, 100
   suckers, 62
Lilium, 40, 64, 138–9
   axillary bulbs, 151
   bulbils, 139
   division, 139
   scales, 138–9, 151
   seed, 138
Lily, 34, 63–64, 138–9
   African, 122, 159
   African corn, 138
   American wood, 125, 140
   arum, 166
   bugle, 140
   day, 124
   giant, 136
   golden-rayed, 64
   Guernsey, 139, 164
   Kaffir, 140
   madonna, 139
   -of-the-field, 140
   -of-the-valley, 123
   Peruvian, 122
   plantain, 124
   St. Bernard's, 122, 136
   St. Bruno, 139
   Scarborough, 140, 166
   spire, 137
   tiger, 64, 139

toad, 125
   water, 128
Lime, 43
Lime tree, 118
Limnanthus, 42, 142
Linaria, 42, 131, 142
Linden, 118
Ling, 107
Linum, 131, 142
Lippia, 112
Liquidambar, 112
Liriodendron tulipifera, 112
Liriope, 124
Lithospermum, 131
Lizard's tail, 128
Loam, 52
Lobelia, 56, 128
Loganberries, 148
   cuttings, 81, 83
   layering, 100
   mist propagation, 71
Lonicera, 74, 112
Loosestrife, 124, 128
   purple, 124
Love-in-a-mist, 141, 142
Love-lies-bleeding, 143
Lunaria annua, 145
Lungwort, 132
Lupin, 124, 142
   cuttings, 75
   mist propagation, 71
   seed, 34, 38, 44
   tree, 34
Lupinus, 124, 142
Lychnis, 34, 124, 131, 142
Lycium, 112
Lysichitum, 124
Lysimachia, 124, 128
Lythrum, 124

Maackia, 112
Macleaya, 124
Madwort, 142
Magnolia, 39, 71, 100, 112
   layering, 100
   mist propagation, 71
   seed, 39
Mahonia, 112
Maidenhair tree, 110
Malcomia maritima, 142
Malling stocks, 85, 149, 150
Mallow, 124, 142
   Indian, 105, 159
   Jew's, 17, 111
   rose, 110–11, 163
   tree, 111
Malope trifida, 142
Malus, 91, 93, 112
Malva, 124
Manure, 42
Maple, 105
Maranta, 164
Marcotting, 104
Marguerite, 161
Marianna rootstock, 150
Marigold, 141
   African, 142

corn, 142
   French, 142
   marsh, 128
   pot, 141, 142
Marjoram, 132, 156
Marrow, 157
   germination, 37
   seed, 33, 34, 35, 51
Marsilea, 128
Masterwort, 123
Matricaria maritima, 142
Matthiola incana, 143, 145
Maurandia, 164
Mayweed, double, 142
Mazus, 132
Meadowsweet, 135
Meconopsis, 39, 132
Medicago echinus, 142
Medinilla, 164
Medlar, 112
Melon, 51, 158
Mentha, 128, 132, 156
Metzelia lindleyi, 142
Menyanthes trifoliata, 128
Merodon equestris, 170
Mertensia, 82, 132
Mesembryanthemum, 143
Mespilus, 100, 112
Metasequoia, 112
Mezereon, 109
Mice, 19
Mignonette, 142
Milkweed, 122, 160
Milkwort, 132
Millipedes, 168
Mimosa, 105, 159
Mimulus, 128, 132, 142
Mind-your-own-business, 165
Mint, 132, 156, 170
   water, 128
Mist propagation, 14, 69–72
Mistletoe, 118
Moluccella, 142
Monarda, 124
Monkey flower, 128
Monkey puzzle, 106
Monkshood, 122
Monocotyledons, 12–13
Monstera, 164
Montbretia, 123, 137
Moonwort, 135
Morina, 124
Morisia, 82, 132
Morning glory, 142, 143, 163
Morus, 112
Moss, creeping, 165
Mother of thousands, 165
Mould, 168, 170
Mouse ear, 131
Mulberry, 73, 107, 112
Mulching, 49
Mullein, 125
Musa, 164
Muscari, 139
Musk, 132
Mustard, 37, 38
Mutation, 29, 30

Myosotidium, 124
Myosotis, 128, 132, 145
Myrobalan, 114
Myrobalan B rootstock, 149
Myrtle, 112
  sand, 111
Myrtus, 112

Nandina, 112
Naphthalene acetic acid, 68
Narcissus, 63, 65, 139, 170
Nasturtium, 142
Nectarine, 153
Neillia, 112
Nemesia strumosa, 143
Nemophila menziesii, 142
Neoregelia, 164
Nepeta, 124
Nerine, 139, 164
Nerium, 164
Nettle, dead, 124
  flame, 161
Nicandra physaloides, 142
Nicotiana, 39, 143
Nierembergia, 132
Nigella, 39, 141, 142
Nitrogen, 42
Nomocharis, 139
Nothofagus, 100, 112
Nuphar, 128
Nymphaea, 128

Oak, 114
Oenothera, 34, 124, 132, 145
Offsets, 61
Oleander, 164
Olearia, 112
Oleaster, 109
Omphalodes, 132
Onion, 156
  diseases, 167, 168
  pests, 169
  seed sowing, 44, 51, 56, 58
  potato, 156
  seed saving, 33
Onopordon, 124
Onosma, 132
Orange blossom, Mexican, 108
Orange, mock, 17, 113
  Californian, 107
Orchid, 38, 164
  poor man's, 165
Origanum, 132, 156
Ornithogalum, 139
Orontium, 128
Orthocide, 168
Osmanthus, 112
× Osmarea, 112
Osteomeles, 112
Othonnopsis, 124
Ourisia, 132
Oxalis, 132, 139
Oxydendrum, 112

Pachysandra, 112
Pachystachys, 164
Paeonia, 40, 124

Palm, 39, 164
Pampas grass, 123
Pancratium, 164
Pandanus, 164
Pansy, 145
Papaver, 42, 124, 142, 145
Paradisea, 139
Paraffin heaters, 22
Paraquat, 48
Parrotia, 112–13
Parsley, 38, 156
Parsnip, 156
  germination, 38
  seed, 33, 35, 44
Parthenocissus, 113
Pasque flower, 132
Passiflora, 113, 164
Passion flower, 113, 164
Paulownia, 113
Pea, Darling River, 165
Peach, 39, 153
  flowering, 114
Pear, 114
  cuttings, 74, 81
  grafting, 85
  rootstocks, 149
Pearlwort, 132
Pearly everlasting, 129
Peas, 156
  disease, 168
  germination, 37
  hybridization, 27–28
  seed, 32, 35, 38, 44
Peat, 52, 55, 78, 97
Pelargonium, 75, 164
  zonal, 76
Peltandra, 128
Penstemon, 34, 75, 132
Peony, 40, 60, 124
  tree, 40
Peperomia, 164
Perennials, 13
  division, 60
  half-hardy, 14–15, 125–8
  herbaceous, 13, 122–5
  woody, 13
Perianth, 30
Periwinkle, 118
Perlite, 78
Pernettya, 80, 113
Perowskia, 113
Pershore rootstock, 150
Pests, 167–70
Petunia, 143
Phacelia campanularia, 142
Pheasant's eye, 122, 129, 142
Phellodendron, 113
Philadelphus, 113
  doubling, 17
  layering, 103
  mist propagation, 71
Phillyrea, 113
Philodendron, 164
Phlomis fruticosa, 113
Phlox, 82, 124, 132, 143
  cuttings, 75, 82
  disease, 170

mist propagation, 71
Phormium, 113
Phosphate, 42
Photinia, 100, 113
Phygelius, 113
Phyllodoce, 113
Phyllostachys, 113
Physalis, 125
Physostegia, 125
Phyteuma, 132
Phytolacca, 125
Phoma lingam, 168
Picea, 14, 89, 113
Pick-a-back plant, 105
Pieris, 100, 113
Pilea, 164
Pimelea, 164
Pimpernel, 129
Pine, 92, 113
  Norfolk Island, 160
  screw, 164
Pineapple, 159
Pink, Chinese, 142
Pinus, 89, 113
Pipings, 76, 77
Piptanthus, 113
Pistil, 31
Pittosporum, 113
Plagianthus, 100
Plane, London, 113
Plants (general)
  adaptations, 11
  characteristics affecting propagation,
    11–12
  hardiness, 14–15
  inheritance, 27–31
  non-seeding, 16–17
  stock, 84
  structure, 12–13, 30–31
Plastic tunnels, 23, 80
Platanus × *acerifolia*, 113
Platycodon, 34, 125, 132
Plectranthus, 164
Plum
  cuttings, 74, 80, 81
  grafting, 91
  rootstocks, 149
  seed, 39
  suckers, 62
Plumbago, 164
Plumeria, 164
Poached egg flower, 142
Poinsettia, 162
Pokeberry, 125
Polemonium, 125, 132
Polianthes, 164
Polka dot plant, 163
Pollen, 30–31
Polyanthus, 145
Polygala, 132
Polygonatum, 125
Polygonum, 113, 132
Pomegranate, 114
Pond-weed, 128
Pontederia cordata, 128
Poplar, 113
  cuttings, 74, 82

grey, 113
root initials, 66
Poppy, 34, 35, 44, 124
  annual, 142
  Californian, 142
  Californian tree, 114
  horned, 123–4
  Iceland, 145
  oriental, 82
  plume, 124
Populus, 113
Potamogeton, 128
Potash, 42
Potatoes, 62–3, 167, 169–70
Potentilla, 71, 113–14, 125, 132
Pots, 25, 27
Peat, 53, 56–7
Potting sheds, 25
Primrose
  Cape, 165
  evening, 124, 132, 145
Primula, 28, 34, 82, 132, 164
  greenhouse, 164
  hybrids, 145
  seed, 34, 35, 39
Privet, 112
  cuttings, 74
  mock, 113
  stock, 86
Propachlor, 48, 154
Propagators, 23–24, 78, 79, 97–98
Prophet flower, 130
Prothalli, 59
Pruning, 153
Prunus, 114, 150, 153
  budding, 93
  grafting, 91
Pseudotsuga, 114
Psila rosae, 169
Pterostyrax, 114
Puccinia menthae, 170
Pulmonaria, 132
Pulsatilla, 82, 132
Punica, 114
Puschkinia, 139
Pyracantha, 80, 114
Pyrethrum, 60, 125
Pyrus, 112, 114
Pythium, 169

Quamash, 136
Quercus, 114
Quince, 74, 85
  Japanese, 107

Rabbits, 19
Radish, 44, 156
Ramonda, 82, 132
Rampion, horned, 132
Ranunculus, 128, 132
  germination, 38
  tuberous-rooted, 139
Raoulia, 132
Raphiolepis, 114
Raspberries, 147–8
  cuttings, 80, 81
  disease, 169, 170

seed, 35
suckers, 62
Rats, 19
Red currants, 147
Red-hot poker, 124
Redwood, Californian, 117
  dawn, 112
  giant, 117
Reedmace, 128
Reinwardtia, 164
Reseda odorata, 142
Reserve garden, 19
Retinosporas, 14
Reversion, 169
Rhamnus, 114
Rhizomes, 62
Rhodanthe, 164–5
Rhododendron, 114
  grafting, 92
  layering, 100, 134
  mist propagation, 69, 71
  rooting, 14, 68
  soil acidity, 43
Rhodohypoxis, 139
Rhoeo, 164
Rhoicissus, 165
Rhubarb, 156
  prickly, 128
Rhus, 12, 82, 114
Ribes, 74, 114, 146–7
Robinia, 12, 114
Rockfoil, 132, 135
Rodgersia, 128
Romneya, 114
Romulea, 139, 140
Rooting, 12,
  bags, 78
  promotion, 66–68
Rosa canina, 85, 119
  hugonis, 121
  laxa, 85, 119
  moyesii, 121
  multiflora, 119
  pendulina, 132
  rugosa, 119
Roscoea, 132
Rose, 119–21
  alpine, 132
  budding, 85, 93, 94, 120–1
  Christmas, 124
  cuttings, 80, 120–1
  dog, 119
  guelder, 118
  layering, 121
  of Sharon, 111, 131
  rock, 131
  stocks, 84, 119
  suckers, 121
  sun, 108
  wild, 119
Rosemary, 114
  marsh, 106, 111
Rosmarinus, 114
Rots, 170
Rubber plant, 104, 162
Rubus, 114, 147–8
Rudbeckia, 60, 125

Rue, 114
  meadow, 125
Runners, 61
Ruscus, 114
Rush, 128
  club, 128
  flowering, 128
Rust, 170
Ruta, 114

Saffron, spring, 136–7
  spring meadow, 136
Sage, 117, 125, 156
  Jerusalem, 113
Sagina, 132
Sagittaria, 128
St. John's wort, marsh, 128
St. Julien A rootstock, 150
Saintpaulia ionantha, 83, 165
Salix, 117
  fragilis, 12
Salsify, 156
Salvia, 117, 125, 142–3, 156, 165
Salpiglossis sinuata, 143
Sambucus, 117
Sand, 52
Sandwort, 130
Sanguinaria, 125
Sansevieria, 83, 152, 165
Santolina, 117
Saponaria, 132, 142
Sarcococca, 117
Satin flower, 135
Saururus, 128
Saxifraga, 132, 135, 165
Scabiosa, 125, 142
Scabious, 60
  giant, 123
  sweet, 142
Schizanthus, 143, 165
Schizophragma, 117
Schizostylis, 140
Scilla, 140
Scindapsus, 165
Scirpus, 128
Scorzonera, 157
Scucellaria, 135
Seakale, 80–81, 157
Sedum, 12, 135
Seed
  annuals, 141
  beds, 42–43
  chitted, 47
  coats, 38–39
  composts, 51–52
  distribution, 11
  dressings, 168
  drills, 44
  drying, 35
  fluid sowing, 46–47
  germination, 16, 35–41
  harvesting, 32–34
  impurities, 36
  pelleted, 45–46
  propagation, 11, 15–17
  quality, 35–36
  selection, 16

slowness to reach maturity, 16
sowing in the open, 42–50, 142
sowing in trays, 144
sowing under glass, 51–59
storage, 34–5
stratification, 39–40
structure, 15, 37
treatment for disease, 167–8
Seedlings
  annuals, 143
  grafts, 86
  pricking out, 57–58
  thinning, 49
  transplanting, 49, 50, 58
Selaginella, 165
Selection, 16
Self-pollination, 29, 31
Sempervivum, 135
Senecio, 117, 142
Senna plant, 161
Sequoia, 117
Sequoiadendron, 117
Setcreasea, 165
Shallots, 156
Shamrock, cape, 132
Shell flower, Indian, 159
Shingle plant, 164
Shoo-fly plant, 142
Shortia, 135
Shrimp plant, 160
Shrubs, 13
  division, 61
  hardwood cuttings, 73–74
  seed, 34
  softwood cuttings, 75
Sidalcea, 34, 125
Silene, 34, 135, 142
Silk oak, 163
Simazine, 48, 154
Sinningia, 163
Sisyrinchium, 135
Sitona lineata, 168
Skimmia, 117
Skull cap, 135
Slipper flower, 130, 160–1
Slugs, 168
Smilacina, 125
Smoulder, 170
Snails, 168
Snapdragon, 129, 143
Sneezeweed, 124
Snowberry, 34, 117
Snowdrop, 137
  tree, 110
Snowflake, 138
Snow-in-summer, 130
Soapwort, 132, 142
Soil
  blocks, 53
  cultivation, 43, 47–49
  for annuals, 141
  heating, 21–22, 23
  for seeds, 42
  sterilization, 25, 52–53
  texture, 19
Solanum, 117, 165
Soldanella, 135

Soleirolia, 165
Solidago, 125
Solomon's seal, 62, 125
Sophora, 117
Sorbus, 91, 117
Sorrel, wood, 139
Sparaxis, 140
Sparmannia, 165
Spartium junceum, 117
Spathiphyllum, 165
Spawn, 63
Spear flower, 160
Speedwell, 125, 135
Sphagnum moss, 78, 104
Spider flower, 143
Spider plant, 161
Spiderwort, 125, 166
Spinach, 157
Spiraea, 117
  alpine, 135
  cuttings, 80
  division, 61
Spores, 18, 58–9
Sports, 29–30
Spot disease, 168
Spring star flower, 138
Spruce, 113
Spurge, 123, 145
Squill, 140
  striped, 139
Stachyurus, 117
Stachys, 125
Stamens, 30
Staphylea, 117
Star of Bethlehem, 139
Stem cuttings, 14, 66–80
Stephanandra, 117
Stephanotis, 165
Sterilizers, 25
Sternbergia, 140
Stewartia, 117
Stigma, 31
Stock, Brompton, 145
  double, 17
  ten week, 143
  Virginian, 142
Stokesia, 125
Stonecrop, 135
Stool bed, 99
Stranvaesia, 117
Strawberries, 148
  disease, 18, 169–70
  mist propagation, 71
  mulching, 49
  pests, 170
  runners, 61, 62
  seed, 35
Strawberry tree, 106
Strelitzia, 165
Streptocarpus, 56, 83, 165
Streptosolen jamesonii, 165
Strobilanthes, 165
Style, 31
Styrax, 117
Succulent plants, 160
Suckers, 61–62, 80
Suckering, 86

Sumach, 12, 82, 114
Sunflower, 124, 142
Swainsona, 165
Swede, 33
Sweet corn, 168
Sweet pea, 34, 38, 142
Sweet rocket, 145
Sweet sultan, 141, 142
Sweet William, 34, 145
Swiss cheese plant, 164
Sycamore, 11
Symphoricarpos, 117
Symphytum, 125
Syringa, 91, 100, 117

Tagetes erecta, 142
Tamarix, 73, 74, 117
Taxodium, 117
Taxus, 117
Tecophilaea, 140
Tellima, 125
Temple bells, 165
Teucrium, 117, 135
Thalia, 128
Thalictrum, 125
Thermopsis, 125
Thiram, 167–8
Thistle, carline, 130
  globe, 123
  Scotch, 124
Thrift, 130
  prickly, 129
Throat-wort, 166
Thuja, 118
Thujopsis, 118
Thunbergia, 165
Thyme, 135, 157
Thymus, 135, 157
Tiarella, 125
Tibouchina, 165
Tickseed, 123, 142
Tidy tips, 142
Tiger flower, 140
Tigridia, 140
Tipula paludosa, 168
Tilia, 118
Toadflax, 131, 142
Tobacco plant, 143
Tolmiea, 166
Tomato, 157
  disease, 168
  hybridization, 29
  light requirements, 24
  pricking out, 58
  seed saving, 33, 34, 35
  seed sowing under glass, 51, 56
Torenia, 166
Trachelium caeruleum, 166
Trachelospermum, 118
Tradescantia, 125, 166
Traveller's joy, 108
Trays, 25, 56–57, 72
Treasure flower, 143
Tree of heaven, 82, 105
Trees, 13, 73–74, 105–18
Tricyrtis, 125
Trifolium, 135

Trillium, 125, 140
Tritonia, 140
Trollius, 125
Tropaeolum, 145
Tsuga, 118
Tuberose, 164
Tubers, 62, 136–40
Tulip
  bulbs, 63, 64
  deblossoming, 65
  disease, 170
  seed, 34
  star, 136
  tree, 112
Tulipa, 140
Tunica, 135
Turnip, 157
  disease, 168
  seed, 33, 35, 44
Turtle-head, 123
Twin-leaf, 131
Typha, 128

Ulex, 118
Ulmus, 118
Umbrella plant, 162
Ursinia, 142, 143
Uvularia, 125

Vaccinium, 80, 103, 118
Valerian, 123
Vallota, 140, 166
Vegetables, 51, 32–34, 154–8
Vegetative propagation, 12, 17–18, 60–68
Venidium fastuosum, 143

Venus' fly trap, 162
Verbascum, 125, 145
  cuttings, 82
  seed, 34
Verbena, 75, 135, 142
  annual, 143
  lemon-scented, 112
Vermiculite, 78
Veronica, 40, 60, 125, 135
Vervain, 135, 142
Vetch, kidney, 129
  milk, 130
Viburnum, 71, 74, 83, 118
Vinca, 118
Vine, 118, 153
  cross, 160
  cup and saucer, 161
  staff, 107
Viola, 75, 135, 143
Violet, African, 83, 165
  dog's tooth, 137
  water, 128
Viruses, 169–70
Viscum, 118
Vitis, 118, 153

Wahlenbergia, 135
Wallflower, 34, 35, 143–5
  alpine, 130
  Siberian, 145
Walnut, 14, 92, 111
Wandering Jew, 166
Wandflower, 123
Water plantain, 128
Water plants, 128
Water supply, 19

Watsonia, 140
Wattle, 105, 159
Wax flower, 163
Weed control, 47–49
Weevils, 168
Weigela, 118
Wellingtonia, 117
White currants, 147
Whorl flower, 124
Willow, 38, 66, 73, 117
  crack, 12
  herb, 123, 131
Wilt, 169
Wind flower, 122, 136
Winter sweet, 108
Wireworm, 168–9
Wisteria, 104, 118
Witch hazel, 110
  American, 110
Woodruff, 130
Wormwood, 122

Yarrow, 122, 129
Yellow oxeye, 123
Yew, 117
Yucca, 39, 118, 166

Zantedeschia, 166
Zauschneria, 135
Zea mays, 15
Zebra plant, 160
Zebrina, 166
Zenobia, 118
Zephyr flower, 140
Zephyranthes, 140
Zinnia elegans, 143